# Contents

# Free-Standing Mathematics Units



# FINANCE

Brian Gaulter +
Leslye Buchanan

OXFORD
UNIVERSITY PRESS

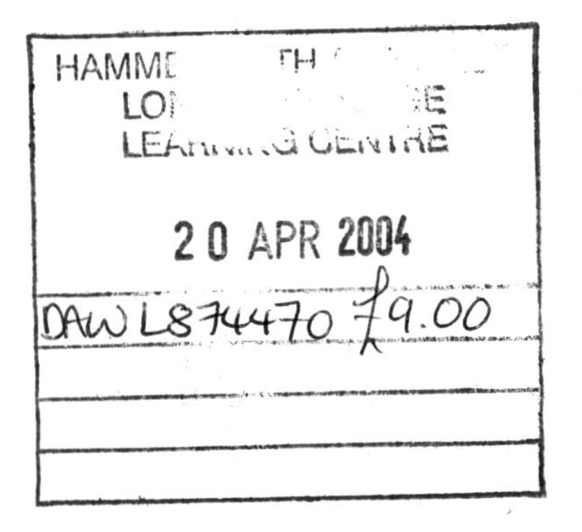



OXFORD
UNIVERSITY PRESS

Great Clarendon Street, Oxford OX2 6DP

Oxford University Press is a department of the University of Oxford. It furthers the University's objective of excellence in research, scholarship, and education by publishing worldwide in

*Oxford New York*
*Athens Auckland Bangkok Bogotá Buenos Aires*
*Calcutta Cape Town Chennai Dar es Salaam*
*Delhi Florence Hong Kong Istanbul Karachi*
*Kuala Lumpur Madrid Melbourne Mexico City*
*Mumbai Nairobi Paris São Paulo Shanghai*
*Singapore Taipei Tokyo Toronto Warsaw*

with associated companies in
*Berlin Ibadan*



First published 2000



British Library Cataloguing in Publication Data

Data available

ISBN 0 19 914797 3

Typeset and illustrated by Tech-Set Limited

Printed and bound in Great Britain

# Introduction

As their name indicates, Free-Standing Mathematics Units, or FSMU, are mathematical courses separate from any other qualification. They have evolved from the Government's decision to encourage more students post-16 to continue with a numerical source.

The units are at three levels:

- Foundation level, broadly equivalent to the lower grades of GCSE,
- Intermediate Level, broadly equivalent to the upper levels of GCSE, and
- Advanced Level, equivalent to the Mathematics found in GCE Advanced or Advanced Supplementary

The units are designed to complement the Key Skills which are used in the Application of Number and to incorporate the majority of these key skills in an integrated post-16 course.

This book is written for students following either:

- Managing Money at Foundation level or
- Calculating Finance at Intermediate Level.

These units cover topics which are comparable to those studied in the Money Management module of the SEG Modular GCSE course. These financial topics are also an ideal study for those following a GNVQ course of a financial nature – for example, Business and Finance – and also for those students following a non-vocational course in a similar area, such as GCE Advanced or Advanced Subsidiary Business Studies.

Candidates studying Managing Money should use the grid opposite, which gives the FSMU level reflected in each topic. In addition to preparing the student for the written examination, the topics contained in this book will enable the student to have the knowledge necessary to complete the portfolio which forms fifty per cent of the final assessment.

Students using textbooks sometimes worry if the solution at the end of the book differs slightly from their own answer. Very often this is caused by a misunderstanding about accuracy. In this book answers involving money should be given to the nearest penny and exact answers should always be given where possible. Where appropriate, other levels of accuracy can be required. As a guide, students should use at least four figures in their working and give answers to three significant figures. The answer must be rounded to the third figure.

We are grateful to Lloyds TSB and the National Westminster Bank for banking documents, HMSO for information from *Social Trends* and the *Annual Abstract of Statistics*, and to the Southern Examining Group for permission to reproduce questions from past GCSE Modular Mathematics examination papers.

We hope that this book will be of great benefit to lecturers and students alike. If you have any suggestions for enhancing its usefulness in future editions, please contact us via the Oxford office of OUP.

Brian Gaulter and Leslye Buchanan
*Hampshire*, 2000

## FSMU LEVELS MATRIX

| Book Section | Foundation | Intermediate |
|---|---|---|
| 1.1 | ✓ | |
| 1.2 | ✓ | |
| 1.3 | | ✓ |
| 2.1 | ✓ | |
| 2.2 | ✓ | |
| 2.3 | ✓ | |
| 2.4 | ✓ | |
| 3.1 | ✓ | |
| 3.2 | ✓ | |
| 3.3 | ✓ | |
| 3.4 | ✓ | |
| 4.1 | ✓ | |
| 4.2 | ✓ | |
| 4.3 | ✓ | |
| 5.1 | ✓ | |
| 5.2 | ✓ | |
| 5.3 | ✓ | |
| 5.4 | ✓ | |
| 5.5 | ✓ | |
| 6.1 | ✓ | |
| 6.2 | ✓ | |
| 6.3 | ✓ | |
| 6.4 | ✓ | |
| 6.5 | | ✓ |
| 7.1 | ✓ | |
| 8.1 | ✓ | |
| 8.2 | | ✓ |
| 9.1 | ✓ | |
| 9.2 | ✓ | |
| 10.1 | ✓ | |
| 10.2 | ✓ | |
| 10.3 | | ✓ |
| 11.1 | | ✓ |
| 11.2 | | ✓ |

| Book Section | Foundation | Intermediate |
|---|---|---|
| 12.1 | | ✓ |
| 12.2 | | ✓ |
| 12.3 | | ✓ |
| 12.4 | | ✓ |
| 13.1 | ✓ | |
| 13.2 | ✓ | |
| 13.3 | ✓ | |
| 14.1 | | ✓ |
| 14.2 | | ✓ |
| 14.3 | | ✓ |
| 14.4 | | ✓ |
| 15 | ✓ | |
| 16.1 | ✓ | |
| 16.2 | ✓ | |
| 16.3 | ✓ | |
| 16.4 | ✓ | |
| 17.1 | ✓ | |
| 17.2 | ✓ | |
| 17.3 | ✓ | |
| 17.4 | ✓ | |
| 17.5 | ✓ | |
| 17.6 | ✓ | |
| 18.1 | ✓ | |
| 18.2 | ✓ | |
| 18.3 | ✓ | |
| 19.1 | | ✓ |
| 19.2 | | ✓ |
| 20.1 | | ✓ |
| 20.2 | | ✓ |
| 20.3 | | ✓ |
| 21.1 | | ✓ |
| 21.2 | | ✓ |
| 21.3 | | ✓ |

# 01 Basic Numeracy

Even in prehistoric times the basis of arithmetic was being developed. The earliest number system was one, two, many. When races began to settle and become farmers, craftsmen and traders, there was a need for more sophisticated number systems for counting, recording and calculation. As different civilisations evolved, they developed different number systems. Our system is based on the number 10. Other civilisations used different systems based on the number 60 (Babylonia), 20 (Mayan) or 5. The Romans used letters to represent numbers (e.g. I, V, X, C for 1, 5, 10, 100).

When trade and communication spread beyond the local area to the rest of the country and abroad, it became essential to have a common system of numbers which was efficient.

The **decimal** system, based on the number 10, was eventually accepted as the most convenient. Its advantages are that any number can be written using only ten symbols, 0, 1, 2, 3, 4, 5, 6, 7, 8, 9 (called **digits**) and **place value**.

## 1.1 Integers and decimal fractions

### Place value

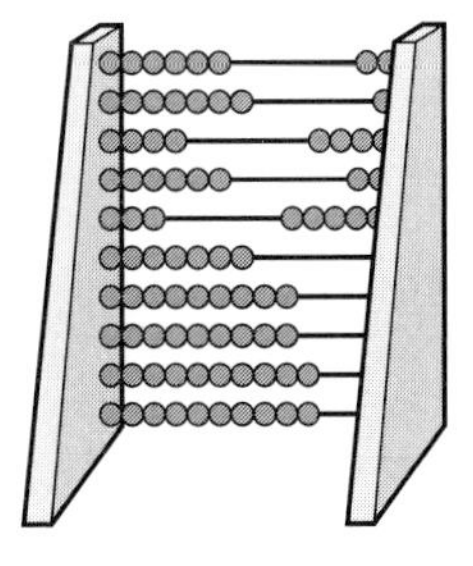

An abacus is a frame with beads sliding on wires, which was used as a counting aid before the adoption of the ten digits. It is still used for this purpose in parts of Asia.

Beads on the first wire have the same value of 1 unit, but a bead on the second wire has a value which is the total of the beads on the first wire, i.e. 10. Each bead on the third wire then has the value $10 \times 10 = 100$, and so on.

The **decimal system** works in the same way, but with digits written in columns instead of beads on wires. Working from right to left, each column is worth 10 times the previous column:

$$15 = 5 \times \mathbf{1} + 1 \times \mathbf{10} = \text{fifteen}$$

$$352 = 2 \times \mathbf{1} + 5 \times \mathbf{10} + 3 \times \mathbf{100} = \text{three hundred and fifty two}$$

$$1234 = 4 \times \mathbf{1} + 3 \times \mathbf{10} + 2 \times \mathbf{100} + 1 \times \mathbf{1000}$$
$$= \text{one thousand, two hundred and thirty four}$$

If there is no digit in a particular column, the place is filled by **0**, which is called a **place-holder**.

$$105 = 5 \times \mathbf{1} + 1 \times \mathbf{100} = \text{one hundred and five}$$

$$150 = 5 \times \mathbf{10} + 1 \times \mathbf{100} = \text{one hundred and fifty}$$

***Note.*** Whole numbers are also called **integers**.

Similarly, working from left to right, each column is worth one tenth of the previous column.

Each number in the column to the right of the units is worth $1 \div 10$, i.e. one tenth, which is a **decimal fraction**.

To separate whole numbers from decimal fractions a **decimal point** is used:

$0.1 = \frac{1}{10}$ = one tenth
$0.01 = \frac{1}{100}$ = one hundredth
$0.001 = \frac{1}{1000}$ = one thousandth

Multiplication and division by 10, 100, 1000, etc., is very easy in the decimal system.

**EXAMPLE 1**

Multiply 34 by: **a** 10 **b** 1000

**a** When a number is multiplied by 10, each digit becomes worth 10 times more, i.e. moves **1 column** to the **left**.
The empty space in the units column is filled by a zero:
$34 \times 10 = 340$

**b** When a number is multiplied by 1000, each digit becomes worth $10 \times 10 \times 10$ more, i.e. moves **3 columns** to the **left**.
The three empty spaces in the units, tens and hundreds columns are filled by zeros:
$34 \times 1000 = 34000$

With numbers of more than four digits, it is usual to leave a small gap after each group of three digits, e.g. 34 000.

**EXAMPLE 2**

Divide 2030 by: **a** 10 **b** 100

**a** When a number is divided by 10, each digit becomes worth 10 times less, i.e. moves **1 column** to the **right**.
The zero in the units column is no longer needed:
$2030 \div 10 = 203$

**b** When a number is divided by 100, each digit is worth $10 \times 10$ times less, i.e. moves **2 columns** to the **right**.

The 3 in the tens column will move into the **tenths column**:
$2030 \div 100 = 20.30$ or $20.3$

## EXERCISE 1.1

**1** What value do the following digits have in the given number?

**a** 5 in 250
**b** 7 in 1725
**c** 3 in 3012
**d** 4 in 624
**e** 6 in 12.6
**f** 2 in 34.529

**2** Write the following numbers in order of size, smallest first:

**a** 54 17 45 10 21 86
**b** 104 23 14 230 32 203
**c** 60 6 600 61 610 601
**d** 99 101 11 110 1001 999
**e** 399 400 300 297 420

3 What is the largest number which can be made from the digits 1, 7, 5 and 0?

4 What is the smallest number which can be made from the digits 3, 9, 6 and 2?

5 Find the answers to the following multiplications:

| | | |
|---|---|---|
| a $43 \times 10$ | d $13.2 \times 10$ | g $5.16 \times 100$ |
| b $167 \times 100$ | e $5.8 \times 100$ | h $7.03 \times 1000$ |
| c $20 \times 100$ | f $93.58 \times 10$ | i $0.13 \times 1000$ |

6 Find the answers to the following divisions:

| | | |
|---|---|---|
| a $320 \div 10$ | d $4030 \div 100$ | g $2.7 \div 1000$ |
| b $300 \div 100$ | e $65 \div 100$ | h $0.51 \div 10$ |
| c $4000 \div 1000$ | f $65 \div 1000$ | i $0.02 \div 100$ |

7 What number is formed when:

| | |
|---|---|
| a 1 is added to 99 | c 1 is added to 4099 |
| b 2 is added to 398 | d 2 is added to 6198? |

## Decimals and the four rules

The four rules of arithmetic are **addition**, **subtraction**, **multiplication** and **division**. It is generally easiest to perform calculations involving decimals on a calculator (see Unit 2). However, in simple cases, you should also be able to find an answer without the aid of a calculator.

**EXAMPLE 1**

a Add 34.5, 9.7 and 56.12.

b Subtract 91.72 from 164.6.

First, write the numbers in a column so that the decimal points are in line, then each digit is in its correct place.

a
$$\begin{array}{r} 34.5 \\ 9.7 \\ 56.12 \\ \hline 100.32 \\ \hline \end{array}$$

b
$$\begin{array}{r} 164.60 \\ 91.72 \\ \hline 72.88 \\ \hline \end{array}$$

**EXAMPLE 2**

a Multiply 271.3 by 9.

b Divide 384.8 by 7.

a
$$\begin{array}{r} 271.3 \\ 9 \\ \hline 2441.7 \\ \hline \end{array}$$

b
$$\begin{array}{r} 54.9 \\ 7\overline{)384.3} \end{array}$$

**EXERCISE 1.2**

Calculate, without the aid of a calculator:

| | | |
|---|---|---|
| 1 $36.2 + 5.7$ | 5 $89.13 - 72.42$ | 9 $147.5 \times 12$ |
| 2 $104.9 + 75.4$ | 6 $121.6 - 69.85$ | 10 $54.6 \div 7$ |
| 3 $6.7 + 51.09 + 76.18$ | 7 $24.2 \times 7$ | 11 $650.79 \div 9$ |
| 4 $72.9 - 56.2$ | 8 $78.04 \times 9$ | 12 $526.46 \div 11$ |

## Combining the four operations

Does $3 + 4 \times 2 = 7 \times 2 = 4$
or
does $3 + 4 \times 2 = 3 + 8 = 11$?

When more than one **operation** is used in a calculation there has to be an agreed order for combining the numbers.

The order used in the calculation is:

(i) brackets
(ii) multiplications and divisions
(iii) additions and subtractions

$\therefore \quad 3 + 4 \times 2 = 3 + 8 = 11$

**EXAMPLE**

Find: **a** $7 + 3 \times 6 - 1$ **b** $(7 + 3) \times 6 - 1$ **c** $7 + 3 \times (6 - 1)$

a $7 + 3 \times 6 - 1 = 7 + 18 - 1 = 24$

b $(7 + 3) \times 6 - 1 = 10 \times 6 - 1 = 60 - 1 = 59$

c $7 + 3 \times (6 - 1) = 7 + 3 \times 5 = 7 + 15 = 22$

### *EXERCISE 1.3*

The answers to questions 1–10 should be found without the aid of a calculator. (You may use a calculator to check your answers.)

**1** $5 + 7 \times 3$

**2** $(5 + 7) \times 3$

**3** $10 \div (5 - 3)$

**4** $10 \div 5 - 1$

**5** $11 - 15 \div 3 \times 2$

**6** $(21 + 17 - 10) \div 4$

**7** $16 \div 4 - 20 \div 5$

**8** $528 \div (150 - 142)$

**9** $150 - 90 \div 45 - 15$

**10** $(150 - 90) \div (45 - 15)$

**11** An artist buys 9 paint brushes costing 56p each.
How much change will be received from a £10 note?

**12** Canvas costs £3.50 per square metre.
How many square metres can be bought for £14?

**13** An office buys two computers for £889.08 each (including VAT) and a laser printer costing £1643.83. The budget for this expenditure is £3500.
How much money remains after the purchases?

**14** A theatre can seat 564 people in the stalls, 228 in the circle, and 196 in the balcony.

**a** How many people can the theatre seat in total?

Seats in the stalls cost £4.50, circle seats cost £6.50, and balcony seats cost £3.75.

**b** How much will the theatre take in ticket sales if it has a full house?

**15** A freelance typist works at 60 wpm. She charges 0.25p per word.

**a** How long will she take to type a document which is 7620 words long?

**b** How much will she receive for this document?

**16** Packs of 40 nappies cost £6.95.
How many packs can be bought for £50?
How much change will be received?

**17** Mrs Halliday uses her own estate car when she delivers meals on wheels and keeps a record of her mileage.

| Week of 24/1/00 | Mon | Wed | Fri | Sun |
|---|---|---|---|---|
| Mileage | 12.3 | 11.5 | 13.9 | 9.8 |

For the week shown above:

**a** what was her total mileage for the week?

**b** how much does she claim for the week if she claims 27.2p per mile?

**18** The diagram shows an extract from a holiday brochure:

| Hotel | Golden Sands | | Park Royal | | Ocean Lodge | |
|---|---|---|---|---|---|---|
| Dates | 14 days | 21 days | 14 days | 21 days | 14 days | 21 days |
| Mar 2–Mar 29 | 323 | 368 | 337 | 382 | 284 | 321 |
| Mar 30–Apr 26 | 373 | 425 | 387 | 439 | 334 | 377 |
| Apr 27–May 24 | 338 | 385 | 352 | 400 | 299 | 339 |
| May 25–Jun 21 | 367 | 418 | 382 | 434 | 329 | 372 |
| Jun 22–Jul 19 | 399 | 455 | 414 | 470 | 361 | 408 |
| Single room supp. | £2.30 per day | | £2.70 per day | | £2.90 per day | |

Find the cost of a holiday for three adults staying at the Park Royal for 14 days from May 25. The third adult will require a single room.

**19** It costs £44.50 per day to hire a car plus £0.06 per mile travelled.
How much does it cost to hire a car for 3 days to travel 450 miles?

**20** A foundry makes accessories for fireplaces. A pair of brass fire dogs weighs 1.7 kg, a set of fire irons weighs 2.04 kg, and a fire screen weighs 3.65 kg.

**a** What is the total weight of the accessories for one fireplace?

A van carries a load of up to 1000 kg.

**b** How many sets of the above accessories can the van carry?

**21** A machine buts lengths of hollow metal rod for lamps. Each section is 20.6 cm long.

**a** How many sections can be cut from a 200 cm length of rod?

**b** What length of rod will be wasted?

## 1.2 Directed numbers

The negative sign has two distinct uses in mathematics:

(i) as a **subtraction** operation, e.g. $6 - 4 = 2$,
(ii) as a **direction** symbol, e.g. $-7\,°C$.

If we wish to show a temperature which is 7 °C *below* zero, we can write $^{-}7\,°C$ *or* $-7\,°C$.

If a car travels 20 miles in one direction and then 15 miles in the reverse direction, we can write the distances travelled as **$^{+}20$ miles** and **$^{-}15$ miles**.

The numbers $^{-}7$, $^{+}20$ and $^{-}15$ are called **directed numbers**.

On your calculator you will see that there are two keys with negative symbols:

[ − ] for subtraction

[ +/− ] for direction.

Directed numbers can be represented on a horizontal or a vertical number line.

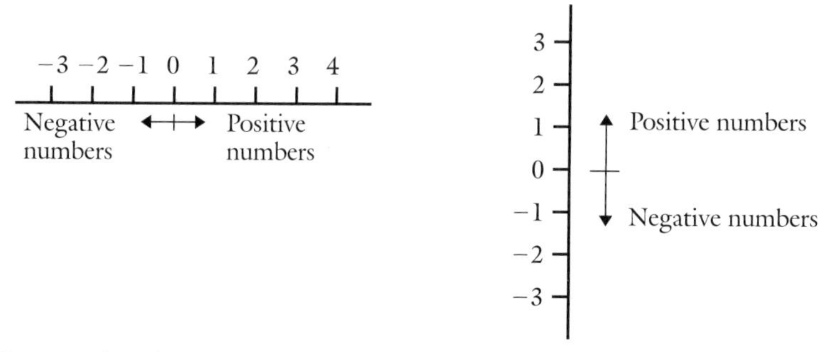


### Addition and subtraction

Think of additions and subtractions of numbers as movements *up* and *down* the vertical number line.

Start at zero, move up 7 and then down 3.
Your position is now 4 *above* zero.

$$\text{i.e.}\quad 7 + (-3) = +4$$

Subtracting two numbers is the same as finding the difference between them:

$5 - (-3)$

is the difference between being 3 *below* zero and being 5 *above* zero on the number line.

From 3 below to 5 above you must move up 8. Therefore

$5 - (-3) = +8$ *or* $5 - (-3) = 5 + 3 = 8$

(Note that $-(-3) = +3$.)

$-4 - (-2)$

is the difference between being 2 below zero and being 4 below zero.

From 2 *below* to 4 *below* you must move DOWN 2, i.e.

$-4 - (-2) = -2$

*or* use the rule $- \times - = +$ to give

$-4 - (-2) = -4 + 2 = -2$

## EXERCISE 1.4

Find the answers to the following, without the aid of a calculator.

**1** a $-5+7$ c $-2+-8$ e $-21+27$ g $-19+19$

b $3+(-7)$ d $16+(-12)$ f $35+5$ h $-35+(-14)$

**2** a $7-8$ c $-9-4$ e $-10+(-7)$ g $25-(-25)$

b $12-(-5)$ d $11-(-9)$ f $-15-(-15)$ h $100-64$.

## Multiplication and division

Going *up* 2 three times would mean you were now 6 *above* zero. So

$$(+2)\times 3 = +6$$

or $2\times 3 = 6$ also $3\times 2 = 6$

Going *down* 2 three times would mean you were now 6 *below* zero. So

$$(-2)\times 3 = -6$$

or $-2\times 3 = -6$ also $3\times -2 = -6$

But what is $(-2)\times(-3)$?

Remember that $-(-6) = +6$

and $(-2)\times 3 = -6 = -(2\times 3)$

$\therefore$ $(-2)\times(-3) = -(2\times -3) = -(-6) = +6$

i.e. the reverse of moving 6 *down* is moving 6 *up*.

The rules for multiplication and division are similar:

**Two numbers with like signs give a positive answer.**
**Two numbers with unlike signs give a negative answer.**

**EXAMPLE 1**

Find the product of $-4\times -7\times -3$

$$(-4\times -7)\times -3 = +28\times -3$$
$$= -84$$

**EXAMPLE 2**

Evaluate $(-12\div 4)\times -6$

$$(-12\div 4)\times -6 = -3\times -6$$
$$= +18$$

## EXERCISE 1.5

Do *not* use a calculator in the following questions:

**1** a $-3\times 6$ d $-7\div -4$ g $8\div\frac{1}{2}\div -2$ j $\dfrac{20\times -6}{-4\times -3}$

b $15\div -5$ e $10\times -3\div 2$ h $(-3+-6)\times -4$

c $-7\times -4$ f $-3\times -3\times -3$ i $-7\div(3-(-4))$

## 1.3 Powers and roots

### Powers

Patterns of dots like this

| • | • • | • • • | • • • • |
|---|---|---|---|
| | • • | • • • | • • • • |
| | | • • • | • • • • |
| | | | • • • • |
| 1 | 4 | 9 | 16 |

form squares, and the numbers 1, 4, 9, 16, ... are called the **square numbers**.

The square numbers are also formed by finding the product of an integer with itself:

$$1 \times 1 = 1$$
$$2 \times 2 = 4$$
$$3 \times 3 = 9$$
$$4 \times 4 = 16, \text{ etc.}$$

The sums of odd numbers also produce the square numbers:

$$1 = 1$$
$$1 + 3 = 4$$
$$1 + 3 + 5 = 9$$
$$1 + 3 + 5 + 7 = 16$$
$$1 + 3 + 5 + 7 + 9 = 25, \text{ etc.}$$

If any number is multiplied by itself, the product is called the square of the number. For example:

the square of 1.2 is 1.44
the square of $\sqrt{2}$ is 2.

The square of 3 or 3 squared $= 3 \times 3 = 3^2$, and the 2 is called the **index**.

Similarly:

the cube of 4 = 4 cubed $= 4 \times 4 \times 4 = 4^3$

and

the fourth power of 2 $= 2 \times 2 \times 2 \times 2 = 2^4$

### Roots

The square of 4 $= 4 \times 4 = 16$.

The number 4, in this example, is called the **square root** of 16, i.e. the number which when squared will give 16.

16 has another square root because $-4 \times -4 = 16$.

Hence the square roots of 16 are +4 and −4.

The symbol $\sqrt{\phantom{x}}$ indicates the positive square root and hence $\sqrt{16} = +4$.

Similarly, the cube root of a number is the number which when multiplied by itself three times equals the given number. For example:

the cube root of 64 or $\sqrt[3]{64} = 4$
(since $4^3 = 4 \times 4 \times 4 = 64$)

## EXERCISE 1.6

1 Write down the first 10 square numbers.

2 Write down the first 6 cube numbers.

3 Find the positive square roots of the following numbers:

a 25 b 121 c 625 d 225 e 196

4 a Find the cube root of (i) 27 (ii) $^-64$ (iii) 729

b Find the fifth root of 32.

c Find the fourth root of (i) 82 (ii) 625

5 Write 25 as the sum of a sequence of odd numbers.

6 What is the third power of 2?

7 What is 3 cubed?

8 Write down the cube root of:

a 8 b 27 c 216 d $\frac{1}{8}$ e $^-64$

9 What is the fourth root of 81?

10 Evaluate $5^2 \times \sqrt{16}$.

# 02 Using a Calculator

## 2.1 Approximations

Most calculators display answers of up to 10 digits.

In most cases, this is too many digits. Therefore, when using a calculator, it is necessary to give an *approximate* answer which contains fewer digits (but which has a sensible degree of accuracy).

There are two methods in common usage:
rounding to a number of **decimal places** (e.g. 2 decimal places or 2 d.p.) and rounding to a number of **significant figures** (e.g. 3 significant figures or 3 s.f.).

The rule for rounding to, for example, 2 decimal places is:

If the digit in the third decimal place is **5 or more**, **round up**, i.e. increase the digit in the second decimal place by 1.

If the digit in the third decimal place is **less than 5**, **round down**, i.e. the digit in the second place remains the same.

**EXAMPLE 1**

Give the following numbers from a calculator display, to 2 d.p.
7.92341, 25.675231, 0.06666. ., 0.9999999. .

| Calculator display | Degree of accuracy required (2 d.p.) | Answer correct to 2 d.p. |
|---|---|---|
| 7.92341 | 7.92\|341 | 7.92 |
| 25.675231 | 25.67\|5231 | 25.68 |
| 0.066666 . | 0.06\|6666 . | 0.07 |
| 0.9999999 . . | 0.99\|99999 . . | 1.00 |

The rules for significant figures are similar, but you need to take care with zeros.

Zeros at the beginning of a decimal number or at the end of an integer are not counted as significant figures, but must be included in the final result. All other zeros are significant.

For example, 70 631.9 given correct to 3 significant figures is 70 600.

The three significant figures are 7, 0 and 6. The last two zeros are not significant (i.e. do not count as fourth and fifth figures), but are essential so that 7 retains its value of 70 thousand and 6 its value of 6 hundred.

**EXAMPLE 2**

Give the following numbers, from a calculator display to 3 s.f.
7.92341, 25.675231, 0.066666. ., 24380.., 0.999999. .

| Calculator display | Degree of accuracy required (3 s.f.) | Answer correct to 3 s.f. |
|---|---|---|
| 7.92341 | 7.92\|341 | 7.92 |
| 25.675231 | 25.6\|75231 | 25.7 |
| 0.066666 . | 0.0666\|66 . . | 0.0667 |
| 24380. | 243\|80. | 24400 |
| 0.9999999. . | 0.999\|999.. | 1.00 |

In the last example, the digit 1 becomes the first significant figure, and the two zeros are the second and third figures.

**EXERCISE 2.1**

Give each of the following numbers to the accuracy requested in brackets:

1 9.736 (3 s.f.)
2 0.36218 (2 d.p.)
3 147.49 (1 d.p.)
4 28.613 (2 s.f.)
5 0.5252 (2 s.f.)
6 4.1983 (2 d.p.)
7 1245.4 (3 s.f.)
8 0.00425 (3 d.p.)
9 273.6 (2 s.f.)
10 459.97314 (1 d.p.)

## 2.2 Estimation

The answer displayed on a calculator will be correct for the values you have entered, but a calculator cannot tell you if you have pressed the wrong key or entered your numbers in the wrong order.

Each number you enter into the calculator should be checked for accuracy and the final answer should be checked by comparing it with an *estimated* answer.

**EXAMPLE**

Estimate the value of $31.41 \times 79.6$.

31.41 is approximately 30
79.6 is approximately 80

An estimated value is therefore $30 \times 80 = 2400$.

If the calculator displays, for example, 25002.36, a mistake has been made with the decimal point, and the answer should read 2500.236.

**EXERCISE 2.2**

1 By rounding all numbers to 1 significant figure, find an estimated value of each calculation:

a $52.2 \times 67.4$
b $6143 \times 0.0381$
c $607 \times 1.86$
d $607 \div 1.86$
e $48.2 \div 0.203$
f $3784 \div 412$
g $\dfrac{520.4 \times 8.065}{99.53}$
h $\dfrac{807}{391.2 \times 0.38}$

2 Find an estimate for each of the following calculations, by choosing an appropiate approximation for each number:

a $82.3 \div 9.1$
b $0.364 \div 6.29$
c $\dfrac{31.73 \times 6.282}{7.918}$

3 By finding an estimate of the answer, state which of the following calculations are obviously incorrect. (Do not use a calculator.)

a $8.14 \times 49.6 = 403.74$
b $23.79 \div 5.57 = 4.27$
c $324 \div 196 \times 0.5 = 226$
d $3.14 \times 9.46^2 = 882.35$
e $23.79 \div 0.213 = 11.169$
f $\dfrac{42.3 \times 3.97}{1635} = 10.27$
g $\sqrt{1640} = 40.5$
h $\sqrt{650} = 80.6$
i $(0.038)^2 = 0.00144$
j $(0.205)^3 = 0.0862$

## 2.3 Degrees of accuracy

Whenever you solve a numerical problem, you must consider the accuracy required in your answer, especially if you cannot obtain an exact answer. In section 2.1, you saw how to correct an answer to a number of significant figures or decimal places. Section 2.3 shows you how to select an appropriate degree of accuracy.

Sometimes the answer to a numerical problem is an integer, and this gives you an exact answer. Fractions are also exact, but their decimal equivalents are often inexact. For example, suppose you were able to buy sixteen pencils for £3 but wanted to buy only one. Each pencil would cost £$\frac{3}{16}$.

As a decimal this would be 18.75 pence, but obviously you cannot pay 18.75 pence for an individual pencil. The shopkeeper would work in the smallest unit of currency, which is one penny. To make sure that she did not lose money she would round up the price to 19p.

**EXAMPLE 1**

A netball club is hiring a number of minibuses to go to an away match. Each minibus can carry 15 passengers and 49 members of the club wish to travel. How many minibuses are needed?

The number of minibuses is $\frac{49}{15} = 3.26667$
3.26667 is rounded up to the nearest integer.

The netball club needs 4 minibuses.

Sometimes you will need to **round down** a mathematical answer as the following example shows.

**EXAMPLE**

Pam changes 9413 French francs into pounds at the rate of 10.9 francs to £1. The bank, not wishing to be generous, decides to round down to the nearest penny the money it will give Pam.

How much does Pam receive?

$$9413 \text{ French francs} = \frac{9413}{10.9} = £863.577\,98$$

$\therefore$ Pam receives £863.57

***EXERCISE 2.3***

1. Twelve pens cost £6.20. What would be charged for one pen?
2. A tin of paint covers $12\,\text{m}^2$.
   How many tins are required to cover an area of $32\,\text{m}^2$ with two coats of paint?
3. Sara gives each of her 35 friends a chocolate biscuit. Each packet of chocolate biscuits contains 8 biscuits. How many packets of chocolate biscuits must Sara open?
4. An office orders plastic sleeves which cost £2.67 for a box of 100.
   How much should be charged:
   **a** for ten **b** for one?
5. A TV room $5.6\,\text{m}^2$ is to be carpeted. The carpet chosen is 3 m wide and is sold in metre lengths. How many metres should be bought?
6. Visitors to a day centre pay £1.70 per week of five days towards the cost of tea and biscuits. How much should someone visiting for one day be charged, if the centre is not to make a loss?
7. Eleven students receive a bill for £120 after an evening out.
   How much should each pay?
8. A booking agency normally sells a block of four tickets for £55.43. It agrees to sell them individually.
   How much should it charge for one ticket if it does not wish to lose money through individual sales?
9. A factory used 26 818 units of electricity at 5.44p per unit.
   What is the cost of the electricity used?
10. In a car factory, pieces of glass fibre 0.9 m are cut from a roll which is 24 m long.
    How many pieces can be cut from one roll?

## 2.4 The keys of a calculator

To use your calculator most effectively, you must become familiar with the keys and their functions.

The booklet that accompanies your calculator will tell you the order in which the keys are used for calculations.

For example, to find $\sqrt[4]{5}$ the keys required on most calculators are

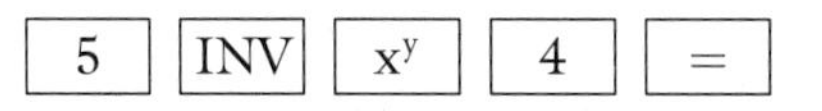

E

...ys are

...ory keys Min MR M+ M− (can you use them correctly?)

...rackets, e.g. $\dfrac{36 \times 14}{21 \times 4}$ is $36 \times 14 \div (21 \times 4)$

or $36 \times 14 \div 21 \div 4$

*not* $36 \times 14 \div 21 \times 4$

**c** The AC key clears the calculator (except the memories) before beginning a new calculation.

**d** The C or CE key is used to correct an entry error (i.e. if the wrong key has been pressed).

The key clears the last entry made (either a figure or an operation), provided it is pressed immediately after the error has been made.

The correct entry can then be made and the sequence continued.

For example, $6 + 3$ C $2 =$ will produce the answer to $6 + 2$.

## EXERCISE 2.4

**1** $386.9 \div 32.87$ (to 1 d.p.)

**2** $2576 \div 0.03741$ (to 3 s.f.)

**3** $\sqrt{\dfrac{79.5}{10.9}}$ (to 3 s.f.)

**4** $0.000356 \times 385.7$ (to 3 d.p.)

**5** $(7.1 + 4.01) \times 8$ (to 3 s.f.)

**6** $\dfrac{1}{0.0345}$ (to 3 s.f.)

**7** $(4.63)^2$ (to 3 s.f.)

**8** $2.1 + 3.41 \times 7.01$ (to 3 s.f.)

**9** $\dfrac{3.8 \times 2.9}{17.1 \times 0.82}$ (to 3 d.p.)

**10** $\frac{1}{2}(246.3 + 1092.8 + 376.4 + 49.8)$

**11** $\dfrac{3.1 \times 15.2}{7.01 \times 8.11}$ (to 3 d.p.)

**12** $\sqrt{\pi \times 38.4}$ (to 2 s.f.)

**13** $\sqrt[3]{4 + 3.7 + 28.1}$ (to 1 d.p.)

# 03 Fractions

## 3.1 Types of fraction

A fraction is a number which can be written as a ratio, with an integer divided by an integer, e.g. $\frac{7}{9}$ or $-\frac{2}{3}$.

If a shape is divided into a number of equal parts and some of those parts are then shaded, the shaded area can be written as a fraction of the whole.

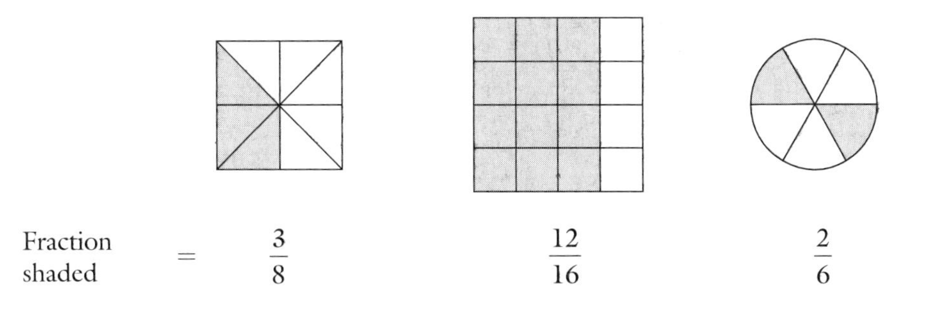


The **denominator** (lower integer) denotes the number of parts into which the shape was divided.

The **numerator** (upper integer) denotes the number of parts which have been shaded.

Fractions which have a smaller numerator than denominator are called **proper** fractions, e.g. $\frac{1}{2}, \frac{5}{12}, -\frac{28}{60}$.

If the numerator is larger than the denominator, the fraction is an **improper** fraction, e.g. $\frac{12}{5}, -\frac{7}{2}, \frac{72}{18}$.

A **mixed number** is a number composed of an integer and a proper fraction, e.g. $3\frac{1}{2}, -24\frac{3}{4}, 4\frac{5}{6}$.

## 3.2 Equivalent fractions

We could consider the larger square above to be divided into four columns instead of sixteen squares.

The shaded area is then $\frac{3}{4}$ of the whole. This means that $\frac{12}{16} = \frac{3}{4}$.

$\frac{12}{16}$ and $\frac{3}{4}$ are called **equivalent fractions** because they have the same value.

**EXAMPLE 1**

Complete $\frac{6}{7} = \frac{?}{28}$ to give equivalent fractions.

The 7 in the denominator must be increased four times to give 28.

The numerator must be treated similarly and be increased four times:

$$\frac{6}{7} = \frac{6 \times 4}{7 \times 4} = \frac{24}{28}$$

**EXAMPLE 2**

Reduce $\frac{32}{72}$ to its lowest terms.

$$\frac{32}{72} = \frac{32 \div 8}{72 \div 8} = \frac{4}{9}$$

(4 and 9 have no common factor therefore the fraction is in its lowest terms.)

**EXAMPLE 3**

Convert $13\frac{2}{5}$ to an improper fraction.

$$\begin{aligned} 13\frac{2}{5} &= 13 + \frac{2}{5} \\ &= (13 \times \frac{5}{5}) + \frac{2}{5} \\ &= \frac{65}{5} + \frac{2}{5} \\ &= \frac{65 + 2}{5} \\ &= \frac{67}{5} \end{aligned}$$

**EXAMPLE 4**

Convert $\frac{141}{22}$ to a mixed number.

$22\overline{)141}$
6 remainder 9

$$\frac{141}{22} = 6\frac{9}{22}$$

## *EXERCISE 3.1*

1 Complete each of the following to give equivalent fractions:

a $\frac{4}{5} = \frac{?}{10}$ c $\frac{3}{4} = \frac{?}{24}$ e $\frac{21}{30} = \frac{?}{10}$

b $\frac{2}{3} = \frac{?}{12}$ d $\frac{1}{4} = \frac{?}{36}$ f $\frac{16}{18} = \frac{8}{?}$

2 Reduce the following fractions to their lowest terms:

a $\frac{10}{15}$ b $\frac{18}{24}$ c $\frac{22}{99}$ d $\frac{21}{33}$

e $\frac{75}{100}$ f $\frac{40}{60}$ g $\frac{30}{65}$ h $\frac{64}{96}$

3 A particular shade of green paint is made by mixing 60 g of blue powder paint with 30 g of yellow powder paint.
What fraction of the mixture is blue paint?

4 A secretary's working day is 8 hours long. On one particular day, $4\frac{1}{2}$ hours were spent typing and $1\frac{1}{2}$ hours on the telephone.
What fraction of the working day was spent:

a typing b telephoning?

5 A dental chart shows that a patient with a full set of 32 teeth has 6 fillings.
What fraction of the teeth are filled?

6 In a group of 240 tourists travelling on a charter flight, 180 requested seats in the non-smoking section.
What fraction of the passengers were in the smoking section?

7 A printing firm produces 350 books in a day. 200 of these are to be exported.
What fraction of the day's production is exported?

8 Convert the following mixed numbers to improper fractions:

a $2\frac{2}{9}$ b $5\frac{1}{6}$ c $4\frac{11}{12}$ d $11\frac{1}{10}$

e $8\frac{3}{4}$ f $12\frac{2}{3}$ g $7\frac{3}{20}$ h $20\frac{3}{7}$

9 Convert the following improper fractions to mixed numbers, in their lowest terms:

a $\frac{9}{7}$ b $\frac{36}{5}$ c $\frac{30}{9}$ d $\frac{61}{8}$

e $\frac{153}{11}$ f $\frac{127}{12}$ g $\frac{132}{15}$ h $\frac{94}{4}$

## 3.3 Operations involving fractions

### Addition and subtraction

Only fractions *of the same type* can be added or subtracted, i.e. they must have the same denominator.

The method for addition or subtraction is:

(i) Find the smallest number which is a multiple of all the denominators.

(ii) Change each fraction to an equivalent fraction with the new denominator.

(iii) Add and/or subtract the fractions.

(iv) If the answer is an improper fraction, convert to a mixed number.

(v) Give the answer in its lowest terms.

**EXAMPLE 1**

Evaluate $\frac{5}{8} - \frac{3}{4} + \frac{1}{5}$

40 is the smallest number that is a multiple of all the denominators.

The sum in equivalent fractions is $\frac{5 \times 5}{8 \times 5} - \frac{3 \times 10}{4 \times 10} + \frac{1 \times 8}{5 \times 8} = \frac{25}{40} - \frac{30}{40} + \frac{8}{40}$

$= \frac{3}{40}$ which is a fraction in its lowest terms.

**EXERCISE 3.2**

Evaluate the following:

1 $\frac{2}{5} + \frac{3}{4}$ 2 $\frac{7}{8} - \frac{5}{6}$ 3 $\frac{5}{12} + \frac{1}{4}$ 4 $1\frac{1}{2} - \frac{3}{4}$

5 A clear glaze for pottery is made by mixing feldspar, flint, whiting and china clay.
One half of the mix is feldspar and one fifth is china clay. Flint and whiting are mixed in equal amounts.
What fraction of the mix is flint?

6 **a** Photographic prints $4\frac{1}{2}$ in wide by $3\frac{1}{2}$ in are to be mounted in an album which has pages $9\frac{3}{4}$ in wide by $13\frac{1}{4}$ in.
How many prints can be mounted on one page?

**b** The photographs are to be equally spaced on the page. What width are the margins:

**a** across the page **b** down the page?

7 Three quarters of an office's stationary budget is spent on paper, one sixth on envelopes, and the remainder on miscellaneous items.
What fraction is spent on miscellaneous items?

8 To encourage customers to pay their bills, a firm gives a discount of $\frac{1}{20}$ of the bill if it is paid on time and a further discount of $\frac{1}{12}$ of the bill for early payment.
What fraction of the bill is deducted for early payment?

9 A dietician, advising clients on suitable diets, found that $\frac{3}{4}$ of her clients were overweight, $\frac{1}{6}$ suffered from arthritis and the remainder from coeliac disease.

**a** What fraction were coeliac sufferers?

**b** If her clients numbered 36 at the time, how many needed a gluten-free diet?

**10** A cold remedy is sold as a powder in sachets. $\frac{4}{5}$ of each powder is aspirin and $\frac{2}{25}$ is ascorbic acid. The remainder is caffeine.

  **a** What fraction is caffeine?

  **b** How much of each ingredient does 500 mg of powder contain?

**11** The proprietor of a B & B establishment mixes his own breakfast cereal. This consists of $2\frac{1}{2}$ cups of oats, $\frac{1}{4}$ of a cup of wheat germ, and $\frac{2}{3}$ of a cup of raisins and nuts.
What is the total number of cups in this mixture?

**12** During a three-day holiday break, $1\frac{1}{2}$ in of rain fell on the first day, $1\frac{3}{4}$ in on the second day and $3\frac{3}{4}$ in on the third day.
How much rain fell altogether?

**13** An axle of diameter $2\frac{3}{4}$ inches is fitted into the centre of the hub of a wheel which has a diameter of $3\frac{1}{8}$ inches.
How much clearance is there between the axle and the inside of the hub on each side?

**14** In a self-assembly unit, a wood top, $\frac{5}{8}$ in thick, is screwed to a metal frame $1\frac{1}{4}$ in thick.
What is the maximum length of screw that can be used?

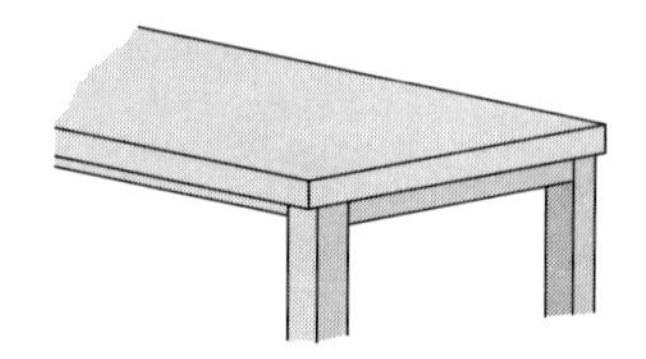

## 3.4 The conversion between fractions and decimal fractions

The fraction $\dfrac{7}{8}$ may be stated as $7 \div 8$.

Using a calculator, $7 \div 8 = 0.875$, and this is the **decimal fraction** which is equivalent to $\dfrac{7}{8}$.

### EXAMPLE 1

Convert $3\dfrac{5}{6}$ to a decimal.

The integer part of the mixed number remains the same. Only the fractional part needs to be converted.

On the calculator $5 \div 6 = 0.833\,3333$ which is a recurring decimal.

$$3\frac{5}{6} = 3.8\dot{3} \text{ or } 3.83 \text{ (to 3 s.f.)}$$

All fractions convert to either a terminating or a recurring decimal. The dot above the 3 indicates that the 3 is a recurring decimal.

### EXAMPLE 2

Convert 0.35 to a fraction in its lowest terms.

$0.35 = \dfrac{35}{100} = \dfrac{7}{20}$ (dividing numerator and denominator by 5)

## EXERCISE 3.3

1 Write down the shaded area as (i) a fraction, (ii) a decimal fraction of the whole area.

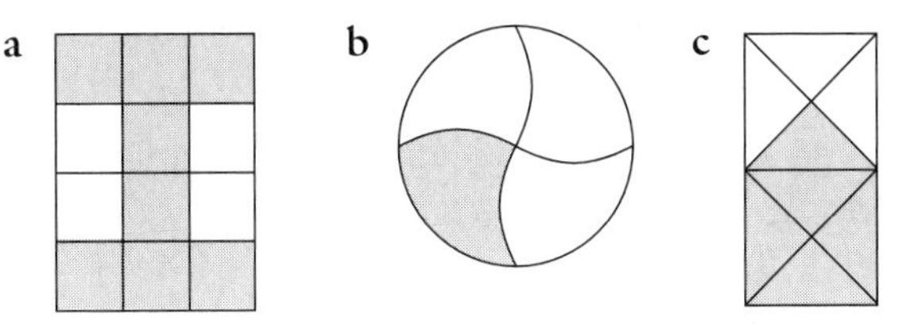


2 Convert the following fractions to decimals:

a $\frac{1}{10}$ b $\frac{1}{2}$ c $\frac{3}{4}$ d $1\frac{9}{20}$

e $4\frac{21}{25}$ f $2\frac{5}{6}$ g $7\frac{4}{9}$ h $3\frac{1}{7}$

3 Convert the following decimals to fractions:

a 0.5 f 2.8

b 0.25 g 3.6

c $1.\dot{6}$ h 2.15

d $1.\dot{3}$ i 0.125

e 1.3 j 0.375

4 Investigate the relationships between the fractional and decimal forms of:

a Halves, quarters and eighths

b thirds, sixths and ninths.

5 Describe a quick method of converting to decimals:

a tenths b fifths c hundredths

d twentieths e twenty-fifths.

In questions 6–11, where appropriate, write the answer:

(i) as a mixed number in its lowest terms

(ii) as a decimal correct to 2 d.p.

6 A designer draws a sketch for a new design of car. The length of the car on the sketch is $4\frac{1}{2}$ inches. The actual length of the car is 153 inches. How many times larger is the actual car than the sketch?

7 a A chemist sells toothpaste in two sizes: 75 g and 125 g.

How many times larger is the 125 g tube than the 75 g tube of toothpaste?

b If the cost of the 75 g tube of paste is 48p what should the equivalent price of the 125 g tube be?

8 Traditionally, 1 quire of paper = 24 sheets
1 ream of paper = 20 quires

Nowadays, however, a ream is generally 500 sheets.

a How many extra sheets are there in a ream?

b How many quires are there in a ream?

9 On a hospital ward, 40 minutes is spent every day checking patients' temperatures and blood pressures.

How many hours are spent on this activity in a 7-day week?

10 A tour bus driver has to travel 112 miles on the first leg of the journey. The driver expects to travel at an average speed of 35 mph.

What is his estimate of the time for this part of the journey?

11 A piece of machinery has two interlocking cogs.

Cog A has 35 teeth. Cog B has 20 teeth.

How many turns does Cog B make for each turn of Cog A?

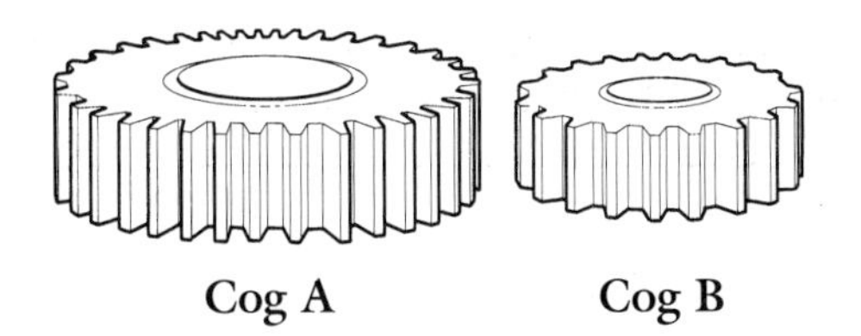

# 04 Ratio and Proportion

## 4.1 Ratio

Carmen's parents give her a weekly allowance of £3.60.
Her younger brother, Leroy, is given an allowance of £1.20.
Carmen receives three times as much allowance as Leroy.
Here is another way of saying the same thing:

The ratio of Carmen's allowance to Leroy's allowance is **3:1**.

Ratios can be written with a colon between the amounts, like this:

$$\text{First quantity} : \text{Second quantity}$$

or as a fraction, like this: $\dfrac{\text{First quantity}}{\text{Second quantity}}$

**EXAMPLE 1**

When Leroy is older, his parents decide that the ratio between his allowance and his sister's should now be 2:3.
If Carmen receives £3.90 per week, how much should they give Leroy?

$$\text{Leroy's allowance} : \text{Carmen's allowance} = 2 : 3 = \frac{2}{3}$$

$$\text{Leroy should receive } \frac{2}{3} \text{ of Carmen's allowance}$$

$$\text{Leroy's allowance} = \frac{2}{3} \times £3.90 = £2.60$$

**EXAMPLE 2**

A large jar of coffee costs £2.38 and a small jar costs 84p.
Express these prices as a ratio in its lowest terms.

Converting both prices to pence gives the ratio

Large jar : small jar = 238 : 84

The ratio can be reduced if 238 and 84 can both be divided by the same number (called a common factor).

The largest factor which is common to 238 and 84 may not be immediately obvious, in which case the reduction to lowest terms can be carried out in stages.

2 is a common factor of 238 and 84. Dividing by 2 reduces the ratio to 119 :42

Possible factors of 42 are 2, 3, 6 and 7.
Only 7 is also a factor of 119.

Dividing by 7 gives the ratio in its lowest terms

$$= 17 : 6$$

**EXERCISE 4.1**

In questions 1–3, write all ratios in their lowest terms.

1 Two brothers have £20 and £24 in their respective savings accounts.
Express these amounts as a ratio.

2 Miss Morgan has £320 in her current account, £400 in her deposit account, and £800 in her savings account.
Express these amounts as a ratio.

3 A pound of grapes costs £1.60 and a pound of pears 72p. Write these prices as a ratio.

4 Write each pair of quantities as a ratio in its lowest terms:

a 60 m, 40 m
b £2, 20p
c 0.6 cm, 0.05 cm
d 0.4 m, 1.6 m
e 1 ft, 9 in
f 15 cm, 10 cm
g 750g, 2 kg
h 39 litres, 26 litres
i 32 mph, 48 mph

5 Complete the following ratios:

a $3:4 = 6:?$
b $18:9 = ?:1$
c $?:1 = 12:10$
d $240:400 = ?:1$
e $20:1 = 64:?$
f $:? = 5:13$

6 Spring bulbs are planted in a border in the ratio of 3 yellow tulips to 2 pink tulips to 5 grape hyacinths.
If 615 yellow tulip bulbs are planted, how many pink tulip bulbs and how many grape hyacinth bulbs are planted?

7 Two terracotta pots have volumes in the ratio of 2 : 9. The smaller can hold 3 kg of peat.
What weight of peat can the larger pot hold?

8 The width of a marigold flowerhead in a photograph is 9 mm. In an enlargement the width is 6 cm.
Write these widths as a ratio.

9 Carl, Stephanie and Joanne deal with 425 clients in a year. The number of clients are in the ratio of 2 : 5 : 10.
a How many clients has Stephanie?
b How many more clients has Joanne than Carl?

10 The cost of a drug to a hospital is 75p and a rest home pays £4 for the same drug.
Write these prices as a ratio.

11 The daily feeds of two new born babies are in the ratio of 3 : 4. The smaller baby needs 21 fl oz per day.
How much milk does the larger baby need?

12 In a leisure centre, the width and height of a locker are 9 inches and 6 feet. Write these measurements as a ratio.

13 Sara, Richard and Francesca hire cars on their holidays. The number of miles travelled by them are in the ratio 4 : 3 : 6. The total mileage travelled is 3120 miles.

a How many miles does Richard travel?
b How many more miles does Francesca travel than Sara?

14 The length of a hacksaw blade is 32 cm and its width is 8 mm. Write these measurements as a ratio.

15 A garage dealing in Ford cars sells Fiestas, Escorts and Mondeos in the ratio of 4 : 5 : 2. In one month 363 of these cars are sold.
How many were Mondeos?

## 4.2 Division in a given ratio

FAIR SHARES FOR ALL!!

This does not necessarily mean equal shares for all.

For example, if three partners invest different amounts of money in a business, they might expect the profits to be shared in proportion to their investment.

**EXAMPLE 1**

Divide £672 between Emily, Faye and Geoff in the ratio 7 : 5 : 9 respectively. How much does each person receive?

*Method*

(i) Find the total number of shares.
(ii) Find the amount of one share.
(iii) Find the amount each receives.

*Calculation*

(i) Total number of shares = $7 + 5 + 9 = 21$

(ii) Amount of one share = $\frac{£672}{21} = £32$

(iii) Emily receives $£32 \times 7 = £224$
Faye receives $£32 \times 5 = £160$
Geoff receives $£32 \times 9 = £288$

(*Check.* $224 + 160 + 288 = 672$.)

**EXAMPLE 2**

Three partners, A, B and C, invest money in a small business. The amounts they invest are £10 000, £12 000 and £6000, respectively.

At the end of the first year of trading the profits from the business are £14 350.

They each receive profits in proportion to their investment.
How much does each partner receive?

The investments are in the ratio 10000 : 12000 : 6000
= 5 : 6 : 3

The total number of shares $= 5 + 6 + 3 = 14$
One share of the profits $= £14350 \div 14 = £1025$
A receives 5 shares $= £1025 \times 5 = £5125$
B receives 6 shares $= £1025 \times 6 = £6150$
C receives 3 shares $= £1025 \times 3 = £3075$

(*Check.* $5125 + 6150 + 3075 = 14350$.)

## EXERCISE 4.2

1 Divide £650 in the ratio 2 : 3.

2 Divide £12 000 in the ratio 1 : 3 : 4.

3 Divide £104 in the ratio 6 : 4 : 3.

4 Mrs Chandra shared £4000 among her three children in the ratio 7 : 5 : 4.
How much did each receive?

5 The Shang Dynasty in China was making bronze artifacts more than three thousand years ago.
The bronze they used was an alloy of copper and zinc in the ratio (by weight) of 17 : 3.
What weights of copper and zinc were used to make a bronze bowl weighing 1.6 kilograms?

6 The number of necklaces, bracelets and earrings made by a jeweller are in the ratio of 4 : 5 : 8.
In one week, she made 68 pieces of jewellery.
How many of the pieces made were bracelets?

7 X and Y invested money in a home computing business. X put in £6000, but Y could only afford £4000. The profits were divided in the same ratio as their investment.

a At the end of the first year the profits were £10 530.
How much did each receive?

b At the end of the second year X's share of the profits was £9456.
How much was the total profit?

c After two years Y increased his investments to £5000. At the end of the year his share of the profits was £8270.
How much did X receive?

8 Four office workers run a pools syndicate and each week pay £4.95, £6.60, £3.30 and £4.95 respectively for their entry. When they win £104 616 they divide the winnings in the ratio of their weekly contribution.
How much does each receive?

9 A drug company representative gives 72 trial samples of a drug to two doctors in the ratio of 3 : 5.
How many samples does each doctor receive?

10 On one day, a hospital casualty department saw 140 accident victims. The ratio of casualties caused by motor accidents, accidents in the home, sporting accidents and others was 4 : 5 : 3 : 2.
How many casualties were the result of motor accidents?

11 a Every summer, Grandma gives her grandchildren money to spend on holiday in the ratio of their ages. When David is 10 years old, Emily is 6 years old.

(i) What is the ratio of their ages, in its simplest form?

(ii) If Grandma gives the children £12 to share, how much does each child receive?

b Next year the children again share £12 between them in the ratio of their ages.
How much (to the nearest 1p) does each receive?

12 The cost of hiring a coach from Bournemouth to London for a day trip was £78. The breakdown of the cost into labour, fuel, overheads and profit was in the ratio of 4 : 4 : 3 : 2.
How much of the cost was for the driver?

13 On one day, a toy manufacturer makes 2072 fashion dolls dressed either in disco wear or in riding gear. The ratio of dolls in disco wear to riding gear is 7 : 1.
How many dolls are wearing disco wear?

14 A company makes 1056 television sets. The number of portable, small screen and large screen models is in the ratio of 4 : 15 : 3.

a How many portable television sets are made?

b How many of the televisions are small screen models?

## 4.3 Direct proportion

If two quantities increase, or decrease, at the same rate, they are said to be in **direct proportion**.

For example, if you double your speed of walking you will travel twice as far in the same period of time.

### EXAMPLE

Mrs Wall usually buys 12 pints of milk each week and pays the milkman £4.92. In a week when she has visitors, she buys 3 extra pints. How much is her milk bill for that week?

12 pints of milk cost £4.92

1 pint of milk costs $\frac{£4.92}{12}$ (=41p)

(12 + 3) pints of milk cost $\frac{£4.92}{12} \times 15$

$= £6.15$

## EXERCISE 4.3

1 Find the cost of 6 lb of apples if 4 lb cost £2.12.

2 The baker's shop sells cheese biscuits for £1.52 per quarter pound.
How much would 7 oz cost? ($\frac{1}{4}$lb=4oz).

3 Pic'n'Mix sweets cost 56p per quarter pound.
How much would you be charged for 9 oz.?

4 A printer charges £1.15 for 25 posters.
How much should be charged for 40 posters?

5 A designer sells coasters in packs of 18 for £19.80. In order to increase sales, the designer decides to offer them also in packs of 4.
How much should he charge for a pack of coasters?

6 A tax adviser charges clients per letter sent. John pays £352 when 16 letters are sent.
How much should Camilla pay if her tax affairs require 11 letters?

7 An accountant charges £369 for 12 hours of her professional services.
How much should she charge for 14 hours?

8 A rest-home needs 24 care assistants when there are 6 residents.
How many assistants does it need when there are 11 residents?

9 Amy burns up 75 calories when she swims for 15 minutes.
How many calories does she use up when she swims for 40 minutes?

10 On an organised hike, it is estimated that hikers take $3\frac{1}{2}$ hours to walk $10\frac{1}{2}$ miles.
How long will it take them to walk 12 miles at the same speed?

11 A timeshare agent is paid for each person he persuades to look at a new development. Alastair sends 32 people and is paid £1312.
How much does Jason receive when he sends 45 people?

12 A carpenter uses 48 screws to fit 4 doors.
How many screws does he need to fit 7 doors?

13 Two spanners are in the ratio of 4 : 11. The smaller is $\frac{1}{4}$".
What size is the larger spanner?

# 05 Percentages

*'Inflation now stands at 2.3%.' 'Ford have given their workforce a 3.5% rise.' 'Unemployment in Winchester is less than 2%.'*

Percentages are a part of our everyday lives. They are often quoted in the media, particularly in connection with money matters.

Percentages often help us make comparisons between numbers, but we must know exactly what a 'percentage' is.

## 5.1 Percentages

A percentage is a fraction with a particular number divided by 100:

$$20\% \text{ means } \frac{20}{100}$$

A decrease of 20% would be a decrease of $\frac{20}{100}$, which is the same thing as a decrease of $\frac{1}{5}$, a fifth.

$$\frac{1}{4} = \frac{1}{4} \times \frac{25}{25} = \frac{25}{100} \text{ which is } 25\%$$

**To convert a fraction to a percentage multiply by 100.**
**To convert a percentage to a fraction, divide by 100.**

**EXAMPLE 1**

Convert **a** $\frac{3}{5}$, **b** $\frac{9}{11}$ to percentages.

**a** $\frac{3}{5}$ as a percentage $= \frac{3}{5} \times 100 = 60\%$

**b** $\frac{9}{11}$ as a percentage $= \frac{9}{11} \times 100 = 81.82\%$ (to 2 d.p.)

**EXAMPLE 2**

Convert 65% to a fraction in its lowest terms.

$$65\% = \frac{65}{100} = \frac{13}{20}$$

**EXERCISE 5.1**

1 Convert the following fractions to percentages:

a $\frac{1}{5}$ b $\frac{1}{8}$ c $\frac{7}{10}$

d $\frac{13}{20}$ e $\frac{2}{3}$ f $\frac{9}{25}$

g $1\frac{3}{4}$ h $2\frac{1}{2}$

2 Convert the following percentages to fractions:

a 60% b 25% c 10%

d 85% e 15% f 130%

g $37\frac{1}{2}\%$ h $33\frac{1}{3}\%$

3 Copy and complete the following table to give each quantity in its fractional, decimal and percentage form.

| | Fraction | Decimal | Percentage |
|---|---|---|---|
| a | $\frac{3}{4}$ | | |
| b | | 0.5 | |
| c | $\frac{1}{8}$ | | |
| d | | | $33\frac{1}{3}$ |
| e | | 0.375 | |
| f | $\frac{7}{10}$ | | |
| g | | | 35 |
| h | | 0.6 | |
| i | $\frac{3}{5}$ | | |
| j | | | 62.5 |

## 5.2 Finding a percentage of an amount

**EXAMPLE 2**

Julian reads in the newspaper that the average pocket money for 12-year-olds has increased nationally by 14% in the last year. Julian's 12-year-old daughter Gilly has been given £1.60 per week for the last two years. How much more per week should he provide for a 14% increase?

$$14\% = \frac{14}{100}$$

$$14\% \text{ of } £1.60 = \frac{14}{100} \times £1.60$$

$$= 0.14 \times £1.60 = 22.4\text{p}$$

Increase in pocket money = 22.4p = 22p (to the nearest 1p)

*Note.* See Section 2.1 for a further explanation of approximations.

**EXERCISE 5.2**

1 Calculate the following percentages to the nearest 1p.

a 10% of £13.75

b 50% of £637.24

c 7% of £316

d 15% of £92.72

e 123% of £4.20

f 121% of £69.80

g 10.80% of £900

h 34.3% of £128.50

## 5.3 Increasing an amount by a given percentage

**EXAMPLE**

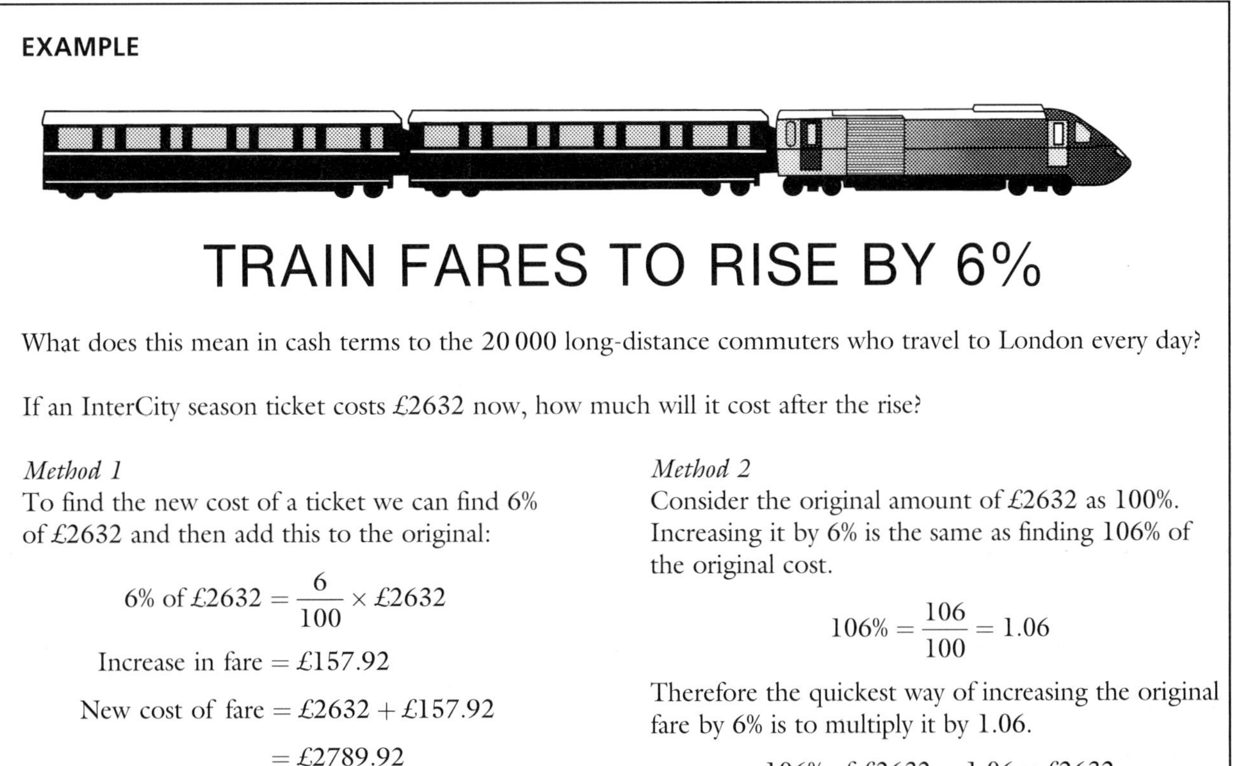


What does this mean in cash terms to the 20 000 long-distance commuters who travel to London every day?

If an InterCity season ticket costs £2632 now, how much will it cost after the rise?

*Method 1*
To find the new cost of a ticket we can find 6% of £2632 and then add this to the original:

$$6\% \text{ of } £2632 = \frac{6}{100} \times £2632$$

$$\text{Increase in fare} = £157.92$$

$$\text{New cost of fare} = £2632 + £157.92$$

$$= £2789.92$$

*Method 2*
Consider the original amount of £2632 as 100%. Increasing it by 6% is the same as finding 106% of the original cost.

$$106\% = \frac{106}{100} = 1.06$$

Therefore the quickest way of increasing the original fare by 6% is to multiply it by 1.06.

$$106\% \text{ of } £2632 = 1.06 \times £2632$$

$$\text{New cost of fare} = £2789.92$$

**EXERCISE 5.3**

Give all answers to the nearest 1p.

1 Increase the following rail fares by 10%.

a £9.20

b £3.70

c £5.00

d £9.81

e £9.13

2 Increase the given amount by the required percentage.

a £72.12 by 50%

b 95p by 10%

c £360 by 120%

d £220 by $6\frac{1}{4}$%

e £124.80 by 25%

f £19.99 by $8\frac{1}{2}$%

## 5.4 Decreasing an amount by a given percentage

**EXAMPLE**

The marked price of this sweater is £24.90.
What is its sale price?

*Method 1*

The reduction = 20% of £24.90

$= \frac{20}{100} \times £24.90$

$= £4.98$

The sale price = £24.90 − £4.98

$= £19.92$

*Method 2*

£24.90 is the equivalent of 100% and so decreasing the price by 20% is equivalent to finding 80% of the original price.

$\frac{80}{100} \times £24.90 = 0.80 \times £24.90$

Sale price = £19.92

**EXERCISE 5.4**

Give all answers to the nearest 1p.

**1** Reduce the following marked prices by 20% to find the sale prices:

**a** £30.00  
**b** £10.50  
**c** £17.60  
**d** 45p  
**e** £12.99

**2** Decrease the given amount by the required percentage:

**a** £54.10 by 8%  
**b** 84p by 30%  
**c** £128 by 60%  
**d** £27.15 by $12\frac{1}{2}$%  
**e** £99.05 by 40%  
**f** £1.62 by 33%

## 5.5 Expressing one quantity as a percentage of another

In a survey of insurance companies it was found that the most common type of car accident was one car running into the back of another.

Out of 35 000 claims, 6280 were for this type of accident.

Information of this type is usually quoted as a percentage.

**EXAMPLE 1**

Find 6280 as a percentage of 35 000.

First express 6280 as a fraction of 35 000:

$$\frac{6280}{35\,000}$$

Then multiply this fraction by 100 to express it as a percentage:

6280 as a percentage of 35 000

$$= \frac{6280}{35\,000} \times 100\%$$

$$= 17.9\% \text{ (3 sf)}$$

This means that almost 18% of car accidents are caused by cars running into the backs of other vehicles.

*Note.* See Unit 2 for a further explanation of significant figures.

**EXAMPLE 2**

A shop buys wallpaper from a wholesaler at £6 per roll and sells it to customers at £8.20 per roll.

What is the percentage increase in price?

The **increase** in price is £8.20 – £6.00 = £2.20

This is $\frac{£2.20}{£6.00}$ as a fraction of the **original** price.

$$\text{Percentage increase} = \frac{£2.20}{£6.00} \times 100\%$$

$$= 36.7\%$$

**EXAMPLE 3**

By what percentage has the marked price of £4.70 been decreased to give a sale price of £3.80?

The **decrease** in price is expressed as a fraction of the **original** price and then multiplied by 100.

$$\text{Decrease in price} = 90\text{p}$$

$$\text{Percentage decrease} = \frac{90\text{p}}{£4.70} \times 100\%$$

$$= \frac{£0.90}{£4.70} \times 100\% \text{ (both quantities must be in the same units)}$$

$$= 19.1\%$$

### EXERCISE 5.5

Give all answers correct to 3 significant figures.

**1** Express the first quantity as a percentage of the second:

**a** 20, 25
**b** 3, 87
**c** 140, 80
**d** 54, 108
**e** 60p, £1.10
**f** £16.25, £12.50

**2** Find the percentage by which the first amount is increased or decreased to give the second amount:

**a** £65, £80
**b** £250, £300
**c** 20p, 95p
**d** £499, £399
**e** £1.23, 67p
**f** £24.50, £138.20

## EXERCISE 5.6

1 A dress made by a Paris Fashion House has a recommended retail price of £870. A London shop advertises it at £710.
By what percentage has the shop reduced the price?
Give your answer to the nearest integer.

2 Linda and Mario are discussing the students at their Art College.

a Linda says that 60% of the students are female and Mario says that there are 52 more females than males.
How many students are there at the college?

b Linda says that $\frac{2}{5}$ of the students attend on Wednesday only in the morning and Mario knows that 30% of the students attend all day on Wednesday.
How many students do not go into college on Wednesday?

3 A watch has a MRRP (maker's recommended retail price) of £32.99, but a jeweller's shop advertises it for £28.99.
By what percentage has the shop reduced the price (to the nearest whole number)?

4 In a town of 25 000 inhabitants, 80% are over 18 years of age.

a How many of the inhabitants are over 18?
Of these, 37% usually shop in the town's supermarket.

b How many shop in the supermarket?

7240 people over 18 living in the town use hypermarkets regularly.

c What percentage of people over 18 shop in hypermarkets?

5 A survey on teenage smoking found that 70% of girls of secondary school age tried smoking and that 36% of those who tried it became addicted.

In a secondary school with 580 female students:

a how many girls would you expect to find had tried smoking?

b how many girls would you expect to find had become addicted to smoking?

6 The 36 residents in a rest-home each pay £231 per week. 28% of the total income of the home is spent on nursing care. The rest-home employs its nurses for a total of 588 hours per week.
What is the hourly rate of pay for each nurse?

7 Restaurants often add a service charge of 12% to your bill. A meal for two costs £28.60.
How much service charge will be added?

8 'Sunny Tours' offers a 5% discount on all holidays booked before 31 December the previous year.

How much will a family of 2 adults and 2 children aged 11 and 15 pay for a holiday whose advertised price is £380 each with a 30% reduction for children under 14 years of age?

9 In the survey of 3500 accidents at work, 17.6% happened on a Friday.
How many of the accidents occurred on a Friday?

10 Mr Robinson invested £32 000, together with money raised from a loan from his bank, in a manufacturing enterprise. He spent 65% on buildings, with an additional 28% on equipment.
How much did he have left to spend on materials?

# 06 Wages and Salaries

## 6.1 Basic pay

All employees receive a wage or salary as payment for their labour.

A **wage** is paid weekly and is calculated on a fixed hourly rate.
A **salary** is paid monthly and is calculated on a fixed annual amount.

Many wage earners are required to work a fixed number of hours in a week, and they are paid for these hours at the basic hourly rate.

**EXAMPLE 1**

Simon works in a hairdressers. His basic pay is £3.75 per hour for a 40-hour week.
Calculate his weekly wage.

$$\text{Weekly wage} = \text{Rate of pay} \times \text{Hours worked}$$
$$= £3.75 \times 40$$
$$= £150$$

**EXAMPLE 2**

Layla's gross weekly wage (i.e. her wage before deductions) is £166.95. She works a 35-hour week.
What is her hourly rate of pay?

$$\text{Hourly rate of pay} = \frac{\text{Weekly wage}}{\text{Hours worked}}$$
$$= \frac{£166.95}{35}$$
$$= £4.77$$

### EXERCISE 6.1

1 Calculate Donna's gross weekly wage if she works for 42 hours per week at a rate of pay of £4.10 per hour.

2 Daniel's yearly salary is £5756.
How much is he paid per month?

3 A basic working week is 36 hours and the weekly wage is £231.48.
What is the basic hourly rate?

4 An employee's gross monthly pay is £965.20.
What is his annual salary?

5 Carrie is an outworker and is paid £3.73 per hour to produce a wall-hanging tapestry.
When Carrie works 36 hours, how much is she paid?

6 Calculate how much Phillipa is paid per year if her monthly pay as a graphics designer is £1975 per month.

7 In one month, Mustafa earns £610 as a secretary.
What is Mustafa's annual income?

8 A shop assistant, Kayley, usually works $48\frac{1}{2}$ hours a week and is paid £221. One week Kayley attends a family wedding and can work only 41 hours.
How much is Kayley paid for that week?

9 As a part-time nurse, Annabel earns £437 per month.
What is Annabel's annual income?

10 Annabel's brother James works as a porter at the same hospital. He is paid £4.52 per hour.
How much is James paid for a week in which he works 43 hours?

11 For her work as a travel courier, Morag is paid £210 per week. She normally works a 38-hour week.
What is Morag's hourly rate of pay?

**12** As a maintenance worker at Stonehenge, Nicholas is paid £3.89 per hour.
How much is Nicholas's gross pay when he works $9\frac{1}{2}$ hours in a day?

**13** Derek is paid £192 when he works a 45-hour week in a factory making parts for a shipyard.
How much is Derek paid per hour?

**14** The basic week in a factory is 35 hours. Find the weekly wage of the following employees whose basic hourly rate of pay is:

**a** a machine operator, basic rate £7.42 per hour

**b** a trainee, basic rate £4.45 per hour

**c** a supervisor, basic rate £12.60 per hour.

## 6.2 Overtime rates

An employee can increase a basic wage by working longer than the basic week, i.e. by doing overtime.
A higher hourly rate is usually paid for these additional hours.
The most common rates are **time and a half** and **double time.**

**EXAMPLE**

Mr Arkwright works a basic 36-hour week for which he is paid a basic rate of £5.84 per hour. In addition, he works 5 hours overtime at time and a half and 3 hours overtime at double time.

Calculate his gross weekly wage.

36 hours basic pay = £5.84 × 36 = £210.24

5 hours overtime at time and a half = (£5.84 × 1.5) × 5 = £43.80

3 hours overtime at double time = (£5.84 × 2) × 3 = £35.04

∴ Gross pay = £289.08

### *EXERCISE 6.2*

**1** Jim Cooper's basic wage is £6.20 per hour, and he works a basic 5-day, 40-hour week. If he works overtime during the week, he is paid at time and a half. Overtime during the weekend is paid at double time.
Calculate Jim's gross wage for the week when he worked five hours overtime during the week and four hours overtime on Saturday.

**2** **a** The basic weekly wage of employees in a small firm is £171 for a 38-hour week.
What is the basic hourly rate?

**b** All overtime is paid at time and a half.
Calculate the number of hours of overtime worked by Ms Wiley during a week when her gross pay was £198.

**3** Jenny, a machinist for a dressmaker, receives £3.91 per hour for her basic pay. One week she works a basic week of 36 hours, together with 8 hours overtime at time and a half.
What is her gross pay for this week?

4 The number of hours worked by an employee is often calculated from a clock card similar to the one shown below. Harold Meyer works a basic eight-hour day, five days a week and his basic hourly rate is £4.15.

| DAY | IN | OUT | IN | OUT | CLOCK HOURS | O/T HOURS |
|---|---|---|---|---|---|---|
| SAT | 0730 | 1200 | 1230 | 1400 | 6 | 6 |
| SUN | 0800 | 1200 | | | | |
| MON | 0730 | 1200 | 1300 | 1700 | | |
| TUES | 0730 | 1200 | 1300 | 1730 | | |
| WED | 0800 | 1200 | 1300 | 1730 | | |
| THURS | 0730 | 1200 | 1245 | 1815 | | |
| FRI | 0730 | 1200 | 1230 | 1600 | | |

Overtime is paid at the following rates:
Time and a quarter for Monday to Friday
Time and a half for Saturday
Double time for Sunday
Calculate Harold's gross pay for this week.

5 Collette normally works 38 hours per week helping to restore 'old masters'. One week she works five hours overtime at double time and receives a gross pay of £269.76.
What is her hourly rate of pay for a weekday evening when Collette is paid time and a half?

6 Janine, a hotel receptionist, is paid £4.70 per hour for her basic 38-hour week. Overtime on Sundays is paid at double time, while other overtime is paid at time and a half.
How much is Janine's gross pay when she works 47 hours in a week, 5 hours of which are on a Sunday?

7 Samantha is an assistant in a personnel section, earning £223.92 when she works a basic 36-hour week. One week she has to produce a summary of last year's appointments and works six hours overtime paid at time and a half.
Calculate Samantha's gross pay for that week.

8 Emily, an assistant in a convalescent home, is paid £3.90 per hour for her basic 38-hour week. During a particularly busy week she does four hours overtime at time and a half and six hours overtime at double time.
Calculate Emily's gross weekly pay.

9 Joanna, a dental assistant, earns £201.60 when she works her basic 36-hour week. In a week after the holidays she also worked 8 hours overtime at time and a half.
Calculate Joanna's gross pay during that week.

10 A swimming pool attendant is paid £3.70 per hour for his basic 36-hour week. He is paid overtime at time and a half.
How much does he earn in a week in which he does 6 hours overtime?

**11** Ruth is a waitress who earns £210.90 when she works a basic 38-hour week. In one week, she had to work five hours overtime (paid at double time) as the restaurant was hosting a wedding reception.
Calculate Ruth's gross pay that week.

**12** The basic working week in a small factory is 35 hours (i.e. 7 hours per day) and the basic rate of pay is £4.98 per hour. The overtime rate is time and a half from Monday to Friday and double time on Saturdays.
The table shown below shows the hours worked by five employees.
For each employee, calculate the gross weekly pay.

| | MON | TUES | WED | THUR | FRI | SAT |
|---|---|---|---|---|---|---|
| Andrews A J | 7 | 7 | 8 | 9 | 8 | 0 |
| Collins F | 8 | 8 | 9 | 9 | 7 | 5 |
| Hammond C | 9 | 9 | 9 | 10 | 10 | 0 |
| Jali Y | 8 | 10 | 11 | 11 | 9 | 4 |
| Longman B H | 9 | 10 | 10 | 9 | 7 | 6 |

**13** Carol works on a production line making televisions. When she works her basic week of 38 hours together with 4 hours overtime at double time, Carol receives a gross pay of £250.70.
What is Carol's basic pay?

## 6.3 Commission

People who are employed as salespersons or representatives and some shop assistants are paid a basic wage plus a percentage of the value of the goods they have sold.

Their basic wage is often small, or non-existent, and the **commission** on their sales forms the largest part or all of their gross pay.

**EXAMPLE**

A salesman earns a basic salary of £690 per month plus a commission of 5% on all sales over £5000.
Find his gross income for a month in which he sold goods to the value of £9400.

He earns commission on (£9400 – £5000) worth of sales

$$\text{Commission} = 5\% \text{ of } (£9400 - £5000)$$
$$= 0.05 \times £4400$$
$$= £220$$

$$\text{Gross salary} = \text{Basic salary} + \text{Commission}$$
$$= £690 + £220$$
$$= £910$$

*EXERCISE 6.3*

1 An estate agent charges a commission of $1\frac{1}{2}\%$ of the value of each house he sells.
How much commission is earned by selling a house for £104 000?

2 Calculate the commission earned by a shop assistant who sold goods to the value of £824 if her rate of commission is 3%.

3 Find the gross monthly pay of Jasmine, who has a basic salary of £855 per month plus a commission of 2% of her monthly sales. She sells £51 410 of goods during the month.

4 A sculptor agrees to pay a studio 6% commission on all sales of her work.

How much commission does the sculptor pay the studio when it sells a sculpture for £8220.

5 Andrina is paid a basic wage of £50 per week by a group of knitters to sell their hand-knitted jumpers. She is also paid 17% commission on all her sales. In one week she sells 160 jumpers at £40 each.
How much is Andrina paid for that week?

6 An insurance representative is paid a commission of 8% on all insurance sold up to a value of £4000 per week. If the value of insurance sold exceeds £4000 per week, he is paid a commission of 18% on the excess.
Calculate his gross pay for the 4 weeks in which his sales were £3900, £4500, £5100 and £2700.

7 Two firms place advertisements for an insurance representative:
Firm A offers an annual salary of £5000 plus a company car (worth £2500 per year) and 4% commission on sales over £200 000 per annum.
Firm B offers an annual salary of £7000 and 3.4% commission on sales over £150 000 per annum.

a Which is the better job if sales of £350 000 per year can be expected?

b If, in a good year, sales rose to £500 000 per year, which job would pay the higher amount and by how much?

8 A nurse at a doctor's surgery is paid a basic wage of £220 per week plus a commission of £2.05 for every vaccination she gives. In one week she givers 31 vaccinations. What is her gross pay for that week?

9 A dentist pays his technician Rachel commission for sealing the teeth of children to prevent tooth decay. Rachel is paid £4 per child up to the first 10 in that week and £3 for each child thereafter. How much commission does Rachel receive in a week in which she seals the teeth of 28 children?

10 Collette, a hairdresser, is paid 12% commission on her work. In one day, her clients pay £500. How much is Collette paid on that day?

11 An agent is paid 5% commission on all tickets sold for shows, and 8% for all coach trips. The agent sells tickets for shows to the value of £900 and £680 worth of coach trips.

Find his commission.

12 A car salesman is paid 21% commission on his weekly sales over £6000. In one particular week he sold two cars for £7520 and £10 640.
What was his commission for that week?

13 In addition to his basic salary of £210 per week, a double-glazing salesman is paid 11% commission on his sales. In one week he negotiates 4 sales of £1250, £180, £3850 and £6100 respectively.

Find his gross income for that week.

## 6.4 Piecework

Some employees, particularly in the manufacturing and building industries, are paid a fixed amount for each article or piece of work they complete. This is known as **piecework**.

They may also receive a small basic wage and, in addition, some are paid a bonus if production exceeds a stipulated amount.

**EXAMPLE**

Workers in a pottery firm who hand-paint the plates are paid a basic weekly wage of £120 and a piecework rate of 80p for every plate over 30 which they paint in a day.

Calculate the weekly wage of an employee whose daily output was as follows:

| | | | | | |
|---|---|---|---|---|---|
| Day 1 | 35 plates | Day 2 | 38 plates | Day 3 | 40 plates |
| Day 4 | 45 plates | Day 5 | 39 plates | | |

No. of plates over 30 painted $= 5 + 8 + 10 + 15 + 9$
$= 47$

Piecework bonus $= 80\text{p} \times 47$
$= £37.60$

Weekly gross pay $= £120 + £37.60$
$= £157.60$

## EXERCISE 6.4

1 A firm employs casual labour to deliver advertising leaflets door to door. The rate of pay is £4.80 for every 100 leaflets delivered. How much is earned by someone who delivers 1230 leaflets?

2 Karl is paid £3.60 per hour plus a bonus of 80 pence for every sack of rubbish collected. In one day Karl collects 30 sacks when he works 9 hours. How much is his gross pay for that day?

3 A seamstress receives a bonus of 57 pence for every 2 fashion waistcoats made in excess of 15 per day. What bonus is received if she makes 30 waistcoats in a day?

4 Mr Ansell, a picture framer, works a basic 40-hour week at £3.78 per hour, plus a bonus of £1.25 for every picture framed. In five successive days he frames 10, 11, 16, 17 and 8 pictures. What is Mr Ansell's gross wage?

5 A freelance computer operator is paid £7 for each page of Basic program which is converted into MAC Basic. In one day she complete 26 pages. How much is she paid for that day?

6 To prepare and complete a tax return, an accountant, Francine, charges a basic £70 plus £4, for every letter sent or received, and £5 for every phone call. In dealing with one client's return, Francine sends 25 letters, receives 18 letters in return and handles 24 phone calls. What is Francine's charge?

7 In addition to his basic salary, a doctor is paid £11.50 for each daytime home visit and £28 for each night call. How much extra does he receive during a weekend when he makes 14 visits of which 5 are during the night?

8 Anya is paid by a charity to deliver lunches to housebound pensioners. Her rate of pay is £1.50 per lunch delivered plus £10 per day. Calculate her pay on a day when she delivers 14 lunches.

9 A timeshare tout is paid £20 for each person she persuades to visit a development. In one week she sends 22 people to the development. How much does the tout receive?

10 Pierre is a guide showing tourists around the Roman amphitheatre in Arles. He is paid 100 francs per group plus a bonus of 10 francs per person. On one day there are 23, 18, 11, 17 and 21 tourists in the six groups he takes on guided tours of the amphitheatre. How much does Pierre earn that day?

11 A bricklayers receives a bonus of 64 pence for every 10 bricks laid in excess of 350 per day. What bonus does he receive if he lays 3070 bricks in a six-day week?

12 The Rapid Fix Tyre Company pays its fitters £4.60 per hour plus 80p for each tyre fitted. In one day Ahmed works 9 hours and fits 71 tyres. What is Ahmed's gross pay for that day?

## 6.5 Deductions from pay

Employees do not usually receive all the money they have earned.

Certain amounts of money are **deducted** from the gross pay and the pay the employee receives is the **net pay** or **take-home pay.**

The main deductions are:

- **Income Tax**
- **National Insurance**
- **Pension** (also called **Superannuation**)

### Income tax (lowest rate)

Income tax is used to finance government expenditure.

It is a tax based on the amount a person earns in a tax year that begins on 6 April and ends on 5 April the following year.

Most people have income tax deducted from their pay before they receive it, by their employer, who then pays the tax to the Government. This method of paying income tax is called **PAYE** (Pay As You Earn).

Certain amounts of each person's income are not taxed. These amounts are called **tax allowances.**

The Tax Office sends the employee and the employer a PAYE Code which tells them the value of the allowances. For example, a tax code of 0450L would be given to a single person who has allowances of £450 × 10 = £4500.

Gross income − Tax allowances = Taxable income

The **lowest** rate of income tax in 1999–2000 was 10p in the pound on taxable income up to £1520.

The Personal tax allowance (in 2000–2001) was £4385.

In addition you may have a tax allowance by reason of expenses connected with your work (e.g. union contributions) or caring for a dependent relative.

**EXAMPLE**

Mr Brown earns £5500 per year, which is his only source of income.

Find **a** his tax allowance, **b** his taxable income,
**c** the tax paid per annum.

**a** Personal tax allowance = £4385

**b** Taxable income = Gross income − Tax allowance
= £5500 − £4385
= £1115

**c** Tax paid $= \frac{10}{100} \times £1115$

$= £111.50$

## EXERCISE 6.5

Assume that the earnings stated in the following questions are the only source of income

1 Find
 (i) the total tax allowance,
 (ii) the taxable income,
 (iii) the yearly tax paid, for:
 a someone earning £4500 per annum
 b someone earning £4950 per annum
 c someone earning £67 per week.

2 Find the yearly income tax payable by Mrs Bailey, who earns £5705 per annum and has a total tax allowance of £4385.

3 Miss Davis earns £6462 per year. She is entitled to the personal allowance plus an allowances of £960 per year for expenses necessary in her work. Calculate her monthly tax bill.

4 Find the yearly income tax payable by:
 a Ulrika, a single art gallery attendant, who earns £5243 per annum and has a total tax allowance of £4385
 b Ashley, a married graphics artist, who earns £105 per week and has a total tax allowance of £4395.

5 A trainee in a design studio is paid £473 per month. His personal allowance is £4385. Calculate the tax paid per month.

6 Zoe's husband is unemployed. Zoe works in a county council office and earns £5750 per year. Calculate her annual tax bill.

7 Jason, an unmarred office junior, earns £530 per month. He has a personal allowance of £5165. Calculate Jason's monthly tax bill.

8 Find:
 (i) the total tax allowance
 (ii) the taxable income
 (iii) the yearly tax paid, for:
 a Sanjay, a married care assistant earning £5200 per annum
 b Francesca, a single nurse, earning £111 per week.

9 Ben, a charity worker, earns £702 per month. He has a personal allowance of £7260. Calculate his monthly tax bill.

10 A travel consultant is paid £460 per month and has a tax-free allowance of £4398. How much tax does he pay per month?

11 A waitress is paid £125 per week. Her personal allowance is £5165. How much tax does she pay per week?

12 Find the annual income tax payable by Malcolm, a single apprentice, working in a steel plant, who earns £6800 per year and has a total tax allowance of £5325.

13 Anna is a trainee machine operator who earns £560 per month. If her personal allowance is £5620, calculate Anna's monthly tax bill.

## Income tax (lowest rate and basic rate)

Most people who pay income tax earn more than £5905, and hence have a taxable income in excess of £1520.

Taxable income between £1520 and £28 400 is taxed at 22p in the pound, i.e. 22%. Since most people pay the majority of their income tax at this rate, the 22p rate is known as the basic rate.

**EXAMPLE**

Mr Jackson is an engineer and earns £21 250 per year. His personal tax allowance is £4557. How much income tax does he pay per year?

Annual taxable income = Annual gross pay – Total allowance
= £21250 – £4557 = £16693

Income taxable at basic rate = Taxable income – £1520 (maximum at lowest rate)
= £15 173

Income tax at lowest rate = 10p × 1520 = £152
Income tax at basic rate = 22p × 15 173 = £3338.06
Total annual income tax = £3490.06

## EXERCISE 6.6

**1** Find the yearly tax paid by:

**a** Andrea, a single girl, earning £12 050 a year

**b** Philip, a single man, earning £182 per week

**c** Sarah and John, a married couple, earning £21 470 between them. Both earn more than £8000.

**2** Jane and Denis live together. Denis earns £13 185 per annum and Jane earns £10 585 per annum.

**a** What was their total tax bill for the year 2000–2001, if they claimed only the standard allowances?

**b** How much tax would they have paid if they had been married?

**3** If, in the next budget, the Chancellor changes the basic rate of tax from 22% to 21%, but wages remain the same, how much less tax would be paid by the employees in question 1 above?

**4** Sally is a design artist earning £27 850 per annum. Her tax allowance is £3445.
How much is Sally's annual tax bill?

**5** Julian, an advertising agent, is paid a basic annual wage of £21 500. She also receives £325 per month in commission.
How much is Julian's monthly tax bill?

**6** Mark, an accountant, earns £31 500 per year.
His tax allowance is £5248.
How much tax does he pay each year?

**7** Find the monthly tax bill payable by Ingrid, who has a company car and is therefore taxed as if she had an additional income of £2700. Ingrid earns £1995 per month and her tax allowance is £4445.

**8** Find the monthly income tax paid by Asif, a staff nurse, who has a tax allowance of £4385 and earns £16 500.

**9** Asif's brother Salman is a porter in the same hospital. His basic wage is £420 per week, and he has a second income of £120 per month.
Salman's tax allowance is £4395.
How much tax does he pay each month?

**10** Marcel, a restaurant owner, has an income of £2310 per month. His tax allowances is £5255.
How much is Marcel's annual tax bill?

**11** Ross, the manager of a health club in an hotel, earns £18 500 per year plus £7450 in bonuses.
His tax allowance is £3855.
How much is Ross's monthly tax bill?

**12** Richard, a window fitter, is paid £285 per week.
He also works on Saturdays and admits to an income of £310 per month from this extra work.
His tax allowance is £4652.
How much tax does Richard pay each month?

**13** As the manager of the 'Exhausts in an Instant' agency, Alf earns a basic salary of £26 150 per year. He is also paid a commission which averages £96 per week. Alf's tax allowance is £4385.
How much tax does Alf pay per year?

## Income tax (higher rate)

Anyone whose taxable income is greater than the amount set by the Chancellor of the Exchequer for basic rate tax payers, has to pay a higher rate of tax on the **excess income.**

For the 2000–2001 tax year the rates payable are:

- 10% lowest rate on a taxable income up to £1520
- 22% basic rate on a taxable income from £1520 up to £28 400
- 40% higher rate on a taxable income over £28 400

**EXAMPLE**

Mrs Carpenter is a director of a small company and earns £37 175 per year. How much income tax does she pay per month?

Her personal allowance = £4385

Annual taxable income = Annual gross pay – Total allowance
= £37 175 – £4385
= £32 790

Income tax at higher rate = Taxable income – £28 400 (maximum at lowest or basic rate)
= £32 790 – £28 400
= £4390

Income tax at lowest rate = 10% of £1520 = £152
Income tax at basic rate = 22% of £26 880 = £5913.60
Income tax at higher rate = 40% of £4390 = £1756

£7821.60

Monthly income tax = £7821.60 ÷ 12 = £651.80

### EXERCISE 6.7

1 Calculate (i) the yearly, (ii) the monthly income tax paid in each of the following cases:

a Anthea Brown earns a salary of £37 000 each year.

b John Knight's annual salary is £38 300. He is married, but his wife is not in paid employment.

c Mr Woolfe's monthly salary is £3010, and his yearly tax allowance is £5060.

2 a David has a full-time job and earns £36 000 per annum. He also has a taxable income of £2000 from shares. His wife Erica has a part-time job which pays £2500 per annum. How much income tax do they pay per year?

b The income from the shares could be transferred to Erica. How much tax will they save if the £2000 is added to Erica's income instead of David's?

3 Paul and his wife Rosemary have a total earned income of £76 600 per annum. Paul's salary is £39 500 per annum. How much income tax do they pay?

4 If, in a budget, the Government decides to have four income tax bands as follows:

10% on the first £3000 of taxable income
21% on the remaining taxable income up to £29 000
35% on the next £5500 of taxable income
45% on the remaining taxable income,

a calculate the yearly income tax which would be payable by Mr Woolfe (see question 1)

b would he benefit under the new system?

## National Insurance

**National Insurance (NI)** is paid by both the **employee** and the **employer.** It helps to pay for:

**National Health Service**
**Statutory sick pay**
**Pensions**
**Unemployment benefits**

The amount paid in National Insurance depends on your gross pay and whether or not you are 'contracted out' of the State Pension Scheme.

Employees who are 'contracted out' pay less National Insurance. On retirement, these employees can claim only the basic State Pension, but they will receive an employment pension.
Employees who are not 'contracted out' receive the basic State Pension plus an earnings-related state pension.

The level of employees' National Insurance contributions was changed in the 2000 budget and from 6 April 2000 is as follows:

| Weekly earnings (£) | Monthly earnings (£) | Rate of National Insurance contribution |
|---|---|---|
| 0–75.99 | 0–329 | Nil |
| 76–535 | 329–2318 | 10% †of earnings in excess of £76 per week (or £329 per month) |
| Above 535 | Above 2318 | 10% of £459 per week (or £1989 per month) |

†8.4% for employees who are contracted out.

**EXAMPLE 1**

Mr Day's annual salary is £15 978. Calculate his monthly National Insurance contribution if he is contracted out.

Mr Day's monthly salary = £15978 ÷ 12 = £1331.50

He earns over £277 per month and will pay the rate of 8.4% (because he is contracted out) on his earnings over £329, i.e. on £1331.50 – £329 = £1002.50.

He pays

8.4% of £1002.50 = £1002.50 × 0.084 = £84.21

Total monthly contribution = £84.21

**EXAMPLE 2**

Mrs Fielding earns a gross weekly wage of £87.90. Calculate her National Insurance contribution.

Mrs Fielding will pay 10% on her earnings over £76, i.e. on £87.90 – £76 = £11.90.

She pays

10% of £11.90 = £11.90 × 0.10 = £1.19

Total weekly contribution = £1.19

It is expected that in the budget in the year 2001 the starting point for paying income tax and for paying National Insurance will become the same. Hence in April 2001 the threshold for paying National Insurance is expected to rise to £87 per week. This will be equal to the anticipated personal tax allowance of £4524 per annum.

### EXERCISE 6.8

1 Using the table on the previous page, calculate the National Insurance contributions for:

a Alice, who earns £112.70 per week

b Charlie, who earns £586.40 per month

c David, who is contracted out and earns £971 per month

d Anne, who is contracted out and earns £428.64 per month

e Maria, who earns £4.30 per hour for a 28-hour week.

2 Mrs Ferry earned £82 per week for her part-time job. Calculate:

a her yearly National Insurance contribution

b her net income for the year.

3 Nadia earns £4.60 per hour for a 40-hour week. Her income tax payment for the week is £29.25. Calculate:

a her gross weekly wage

b her weekly National Insurance contribution

c her net pay for the week.

*Note.* Section 6.5 takes into account the changes announced in the Spring Budget for the financial year 2000–2001. Readers using this book during or after the financial year 2000–2001 should make themselves aware of any further changes introduced by the Chancellor of the Exchequer.

# 07 Travel

## 7.1 Foreign currency

### Currency exchange

If you plan a trip abroad for business or holiday, you must decide the money you will need and the form in which you will take it.

Most people take a limited amount of cash in the currency of each country to be visited. They take the remainder in the more secure form of traveller's cheques. Alternatively, they use credit cards to obtain cash.

In the UK British currency (pounds sterling) is usually exchanged for foreign currency at a bank or travel agency. Exchange rates between one currency and other currencies change frequently and are published daily in newspapers and displayed where money is exchanged.

By consulting the **selling price** you can calculate the amount of foreign currency you will be sold for your pounds and from the **buying price** you can calculate the amount of pounds you will receive in return for your foreign currency.

**EXAMPLE**

Mr Oztürk is to travel on a business trip to Greece. He changes £350 into Greek drachma on a day when the bank selling price is 502.2.

On returning home he changes his remaining 105 500 drachma into sterling. The bank buying price is 528.9 to £1.

Calculate:

a the amount of drachma he receives

b the amount of pounds he receives on his return.

a The bank pays 502.2 drachma for every £1 it buys

For £350 he will receive 502.2 × 350 drachma

= 175 770 drachma

b The bank charges 528.9 drachma for every £1 it sells

For 105 500 drachma he will receive $£\frac{105\,500}{528.9}$

= £199.47

EXERCISE 7.1

1 Using the bank selling price, change:

a £12 to Croatian kunas

b £140 to Latvian lats

c £96.50 to American dollars.

2 Using the bank buying price, calculate the sterling equivalent of:

a 200 Swiss francs

b 11 100 Canadian dollars

c 650 Japanese yen.

| EXCHANGE RATES | | |
|---|---|---|
| | Bank buys | Bank sells |
| Australia | 2.56 | 2.42 |
| Canada | 2.44 | 2.30 |
| Croatia | 12.85 | 11.48 |
| Czech Republic | 58.21 | 55.31 |
| Greece | 528.9 | 502.2 |
| Japan | 170.2 | 160.8 |
| Latvia | 0.971 | 0.916 |
| Lithuania | 6.60 | 6.30 |
| Norway | 8.208 | 7.715 |
| Poland | 6.916 | 6.522 |
| South Africa | 9.67 | 9.12 |
| Slovenia | 330.7 | 300.2 |
| Switzerland | 2.62 | 2.48 |
| Turkey | 837 700 | 787 440 |
| USA | 1.65 | 1.58 |

3 Geraldine and Peter ate a meal in a restaurant while on holiday in Rhodes.
The meal for two cost 4256 drachma. Use the bank selling price to calculate the cost of the meal in pounds.

4 The Andersons spent a holiday touring in Switzerland. While travelling they used 200 litres of petrol which cost 1.04 francs per litre.
The exchange rate was 2.59 Swiss francs to £1.

a How much did the petrol cost them in pounds?

b What was the price per litre of the petrol in pence?

5 Before going on holiday to Eastern Europe, the Williams family changed £600 into Polish zloty.
While in Poland they spent 1840 zloty and then changed their remaining zloty into Latvian lats as they crossed the border.
The exchange rate was 0.1405 lats to 1 zloty.
Calculate:

a the number of zloty they received

b the number of lats they bought.

6 If Mr Oztürk (in the example on the previous page) had postponed his trip to Greece until the following week the bank selling price would have been 508.4 drachma to the pound.
On his return, the buying price would have been 534.8 drachma to the pound.

a How many drachma would he have received?

b How many drachma would he have had left on his return to Britain (assuming that he would have spent the same amount)?

c How many pounds would he have received on his return?

d How much money would he have saved by travelling the following week?

7 a Mr Elton changed £100 into Norwegian krone for a day trip to Norway. How many krone did he receive?

b Unfortunately the excursion was cancelled and so he changed all his krone back to pounds. How many pounds did he receive?

c How much money did he lose because of the cancellation?

## Commission

In practice, the banks also charge commission for each currency exchange.

The commission is £1 on currency exchanges up to the value of £200 and 0.5% of the value above £200.

On traveller's cheques the commission is £2 for up to £200 in value and 1% of the value above £200.

### *EXERCISE 7.2*

Questions 1 and 2 refer to Exercise 7.1.

1 a How much commission did Mr Elton pay when changing pounds to krone?

b How much commission did he pay when changing the krone back to pounds?

c How much did his cancelled trip cost him, including commission charges?

2 a How much commission did the Williams family pay for their zloty?

b How much commission would they have paid if they had taken the £600 in traveller's cheques?

# 08 Value-added Tax

Apart from income tax, the Government also raises money by taxes on goods and services. These are called **indirect taxes** as they are included in the price to the consumer. For example, the prices of alcohol, tobacco and petrol include a high rate of tax.

## 8.1 Prices exclusive of VAT

The main indirect tax is **value-added tax** or **VAT**. This is charged on most goods that we buy and also on services such as house improvements, garage bills, meals in a restaurant.

Certain items are **zero-rated**, i.e. no VAT is charged, and these include most food (but not luxury foods such as ice cream, confectionery and crisps), children's clothes and books.

**The standard rate of VAT is 17.5% (2000–2001 value)**

Prices are sometimes quoted exclusive of VAT. The price you pay will include the VAT and will be 17.5% more than the price quoted.

**EXAMPLE 1**

A garage bill totals £88.70. How much VAT is added to the bill?

$$\text{VAT} = 17.5\% \text{ of } £88.70$$

$$= \frac{17.5}{100} \times £88.70 \quad \text{or} \quad 0.175 \times £88.70$$

$$= £15.52 \text{ (to the nearest 1p)}$$

In the 2000 budget, the rate of VAT on fuel (gas, electricity, etc.) was fixed at 5%.

**EXAMPLE 2**

An electricity bill is £112.32 plus 5% VAT. How much is the VAT?

$$\text{VAT} = 5\% \text{ of } £112.32$$

$$= \frac{5}{100} \times £112.32$$

$$= £5.62 \text{ (to the nearest 1p)}$$

EXERCISE 8.1

The prices of the items in questions **1–5** are quoted exclusive of VAT. Calculate the amount of VAT payable in each case, giving your answer to the nearest 1p.

**1** a typewriter priced at £152

**2** a washing machine priced at £234

**3** a calculator at £9.50

**4** a box of chocolates at £1.44

**5** a necklace at £19.20

**6** A garage bill shows the cost of parts to be £30.90 and labour costs of £58.60. VAT is then added to the bill. How much is the final bill?

**7** A particular calculator is offered for sale through three different catalogues at the following prices (p&p means postage and packing):

Catalogue A £12.60 exclusive of VAT, post free

Catalogue B £13.80 inclusive of VAT, plus 50p p&p

Catalogue C £12.40 exclusive of VAT, plus 40p p&p.

Calculate the price in each case and hence determine which catalogue is offering the best deal.

**8** All prices at a DIY store are shown exclusive of VAT. Mrs Jennings buys eight boxes of tiles, priced at £9.90 per box, a large packet of tile cement at £2.39 and some ready mixed tile grout at £1.38.
Calculate her total bill, including VAT.

**9** The cash and carry store shows prices exclusive of VAT. Mr Patel buys the following goods for his grocery shop: £45.30 worth of food which is zero-rated for tax, £27.80 worth of confectionery and £15.25 worth of toiletries. Calculate the amount of VAT payable and the total amount of his bill, including VAT.

**10** The Government decides to change the rate of VAT payable on goods.
Calculate the amount of VAT payable and the total cost of the following goods, including VAT at the rate given:

**a** a paperback at £4.95, VAT at 12%

**b** a gas bill of £340, VAT at 5%

**c** a personal stereo at £28.90, VAT at 9%

**d** a patio door at £276, VAT at 17%

**e** two garden gnomes at £6.87 each, VAT at 14%.

## 8.2 Prices inclusive of VAT

Most shop prices are quoted inclusive of VAT. An item costing £100 *exclusive* of VAT would have VAT of £17.50 added and its price *inclusive* of VAT would be £117.50. Therefore, for all prices which are inclusive of VAT,

**every £117.50 of the price includes £17.50 of VAT**

That is, if the price exclusive of VAT is 100%, the price inclusive of VAT is 117.5% of the price.

**EXAMPLE**

A reclining chair costs £256.45, including VAT at 17.5%.

Calculate:

**a** the amount of VAT which was added,

**b** the cost of the chair, excluding VAT.

**a** Price including VAT = £256.45 = 117.5% of the price

$$\therefore \quad 1\% \text{ of the price} = \frac{£256.45}{117.5}$$

$$\text{VAT, equivalent to } 17.5\% = \frac{£256.45}{117.5} \times 17.5$$

$$= £256.45 \times \frac{17.5}{117.5}$$

$$= £38.19$$

**b** $$\text{Price excluding VAT, equivalent to } 100\% = \frac{£256.45}{117.5} \times 100$$

$$= £256.45 \times \frac{100}{117.5}$$

$$= £218.26$$

Calculating another way gives:

Price excluding VAT = £256.45 – £38.19 (VAT) = £218.26

**To find the VAT included in a price, divide by 117.5 and multiply by 17.5**

**To find the price excluding the VAT, divide by 117.5 and multiply by 100.**

These calculations only apply to VAT at 17.5%.

For fuel bills where VAT is at 5%:

**To find the VAT included in a price, divide by 105 and multiply by 5.**

**To find the price excluding VAT, divide by 105 and multiply by 100.**

If the VAT rate has changed since 2000, can you work out how to change the figures?

## EXERCISE 8.2

1 A bottle of whisky costs £8.20.
How much of this price is VAT?

2 Mr Lambert buys a motor-cruiser costing £10 710.
What is the price of the motor-cruiser excluding VAT?

3 How much VAT will a shopkeeper pay to the Government on takings of £76.80 for sweets and £43.95 for ice cream?

4 A garage sells 5000 litres of lead replacement petrol at 85.9p per litre and 9000 litres of unleaded petrol at 82.4p per litre.
How much VAT does the Government receive from the sale of this petrol?

5 Mrs Jali received a garage bill totalling £125.20 for parts and labour including VAT.
How much did the garage charge for parts and labour?

6 A school buys a computer which costs £899.90, including VAT. Schools are not required to pay VAT on educational equipment.

**a** How much VAT can the school reclaim?

**b** How much does the school pay for the computer?

7 The items listed below are on sale in a large electrical store. Assume that the Government has suddenly reduced the rate of VAT to 12%. For each item, calculate the amount of VAT paid and the price before VAT was added (give answers to the nearest 1p):

**a** steam iron, £28.99 **b** hair-dryer, £17.99

**c** shaver, £54.99 **d** microwave, £199.99

**e** washing machine, £329.99

**f** vacuum cleaner, £114.99

**g** television, £599.99 **h** can opener, £11.99

**8** The bill for a meal at a restaurant in British Columbia was $63.90, including VAT at the rate of 8%.
How much was the VAT?

**9** John's gas bill is £185.76 inclusive of VAT at 5%. How much was the bill before VAT was added?

**10** Rebecca's electricity bill is £206.54 inclusive of VAT at 5%. How much VAT was charged?

# 09 Profit and Loss

## 9.1 Calculations involving one price

Everyone who is involved with buying and selling goods aims to make a **profit.**

To do so the dealer buys the goods for a certain amount, the **cost price,** and sells the same goods for a higher price, the **selling price.**

The difference between the cost price and the selling price is the profit:

$$\textbf{Profit} = \textbf{Selling price} - \textbf{Cost price}$$

Because it is easier to compare percentages, the **percentage profit** is often required. It is usually calculated on the cost price:

$$\textbf{Profit \%} = \frac{\textbf{Profit}}{\textbf{Cost price}} \times \textbf{100\%}$$

Sometimes the dealer makes a loss, i.e. the goods are sold for less than the cost price.

In this case, the percentage loss is calculated on the difference between selling price and cost price:

$$\textbf{Loss \%} = \frac{\textbf{Loss}}{\textbf{Cost price}} \times \textbf{100\%}$$

**EXAMPLE 1**

A shoe shop buys a particular style of shoe for £22 per pair and sells these shoes to the customer for £29.99 per pair.

**a** What profit does the shop make?

**b** What percentage profit does the shop make?

**a** Profit = Selling price – Cost price
= £29.99 – £22
= £7.99

**b** $\text{Profit \%} = \frac{\text{Profit}}{\text{Cost price}} \times 100\%$

$= \frac{£7.99}{£22} \times 100\%$

$= 36.3\%$

**EXAMPLE 2**

A car is bought for £7290 and sold three months later for £6925.
What percentage loss was made on the deal?

Loss = £7290 – £6925
= £365

$\text{Loss \%} = \frac{£365}{£7290} \times 100\%$

$= 5.0\%$

**EXAMPLE 3**

A supermarket buys cheese for 120p per pound. When the cheese reaches its 'sell by date' the retail price is reduced and the supermarket makes a loss of 10%.
At what price per pound is the cheese now sold?

$$\text{Cost price } (100\%) = 120\text{p}$$

$$10\% \text{ loss} = \frac{10}{100} \times 120\text{p} = 12\text{p}$$

$$\text{Selling price} = 120\text{p} - 12\text{p}$$

$$= 108\text{p per pound}$$

*EXERCISE 9.1*

1 For each cost price (CP) and selling price (SP) given below, find: (i) the profit or loss and (ii) the percentage profit or loss.

a CP = £4, SP = £5

b CP = £20, SP = £27

c CP = 75p, SP = 66p

d CP = £90, SP = £108

e CP = £525, SP = £425

f CP = £36, SP = £39.99

2 A market trader buys kiwi fruit at £9 for a box of 96 and sells them at 8 for £1.
What is his percentage profit?

3 A small greengrocer's shop buys crates containing 24 melons for £36 per crate. The melons are then sold for £1.68 each.

a What profit does the shop make on each crate of melons?

b What is this profit as a percentage of the cost price?

4 A car was bought for £10 565 and resold at a loss of 8%. What was the selling price?

5 A greengrocer buys a 10 lb box of grapes for £7.20. The grapes are sold to customers at £1.08 per lb.
If 2 lb of grapes are unsaleable, what is the greengrocer's percentage profit on the cost price?

6 A retailer buys calculators for £29.20 each plus 17.5% VAT and resells them at £41.98 each.
What is the retailer's profit as a percentage of his outlay?

7 a A house was bought in 1980 for £82 000 and sold 10 years later for £160 000.
What percentage profit was made?

b If the purchasing power in 1990 of a 1980 £1 had eroded to 52p, what was the true profit as a percentage of £82 000?

## 9.2 Calculations involving more than one price

**EXAMPLE**

A hardware shop buys 18 saws for £8.67 each. Thirteen are sold for £10.85, and the remaining saws are sold at the reduced price of £7.99 in the January sale.

a Calculate the overall profit which the shop makes on the 18 saws.

b Calculate the shop's percentage profit.

a Purchase price for 18 saws $= £8.67 \times 18 = £156.06$
Amount received for 13 saws $= £10.85 \times 13 = £141.05$
Amount received for 5 saws $= £7.99 \times 5 = £39.95$

Total amount received $= £141.05 + £39.95 = £181.00$
Overall profit $= £181.00 - £156.06 = £24.94$

b Percentage profit $= \dfrac{\text{Profit}}{\text{Purchase price}} \times 100 = \dfrac{£24.94}{£156.06} \times 100\%$
$= 16.0\%$

## EXERCISE 9.2

1 A chemist's shop buys 25 bottles of a new brand of perfume for £195. To attract sales, the first ten bottles are sold for £8 each. The remaining bottles are all sold at the normal price of £10 each. Calculate:
   a the overall profit which the shop makes on this perfume
   b the percentage profit.

2 An electrical store buys fifty personal stereos for £949.50. Thirty five are sold for £24.99 each. The remaining stereos are all sold at the reduced price of £19.99 each. Calculate:
   a the total amount received for the 50 stereos
   b the overall profit which the store makes on these 50 stereos
   c the store's profit, as a percentage of the purchase price.

3 A grocery store buys 500 tins of tomatoes for 23p per tin. The tins of tomatoes are sold to the public for 29p. Unfortunately, 36 tins have been dented and these tins are sold for 5p less than normal. Assuming all the tins are sold, calculate:
   a the shop's overall profit
   b the shop's percentage profit.

4 A garage sells bunches of fresh flowers which it buys from a local nursery.
The garage pays £1.50 per bunch and sells them for £2.50 per bunch. Any bunches remaining after three days are sold at half price and any remaining after five days are thrown away. On a particular day, the garage buys 12 bunches of flowers. Seven bunches are sold at full price and three are sold at half price. Calculate:
   a the overall profit on these 12 bunches
   b the percentage profit.

5 The clothing department of a large store purchased 100 matching shorts and T-shirts as a special line for the summer season. The purchase prices were £7 for each pair of shorts and £5 for each T-shirt.
The price to the general public was £9.99 for a pair of shorts and £7.99 for a T-shirt. The items could be bought separately.
Eighty pairs of shorts and ninety T-shirts were sold at these prices. Calculate:
   a the total purchase price of 100 shorts and T-shirts
   b the amount the shop received for 80 pairs of shorts and 90 T-shirts.

   The remaining shorts and T-shirts were all sold in the 'end of season' sale for £2 each less than the original selling price.
   c Calculate the amount the shop received for the shorts and T-shirts during the sale.
   d What was the total amount received by the shop for these items?
   e Find the overall profit which the shop made on the shorts and T-shirts and express this as a percentage of the purchase price.

6 Barry buys 200 watches and sells 60% of them for £9 each.
   a How much does he receive?

   Barry then reduces the price of the remaining stock to $\frac{3}{4}$ of the selling price.
   b What is the new price of a watch?

   Barry sells $\frac{3}{5}$ of the remaining stock at this new price. Unfortunately, the remainder are broken and are thrown away.
   c What is the *total* amount of money he received from the sale of the watches?

   Barry paid £820 for the 200 watches.
   d Express the profit made as a percentage of the cost price. Give your answer correct to 2 decimal places. (SEG W93)

# 10 Savings

There are many ways people can choose to invest their money, some of which are discussed later in this unit.

Most people will invest some, if not all, of their money in a building society, post office or bank savings account.

Because they are **lending** their money they will receive **interest** on the money invested.

## 10.1 Simple interest

The initial sum of money invested is called the **principal**. With **simple** interest the amount of interest to be paid is always calculated on the principal, and therefore the interest remains the same every year.

In practice, this will only happen if the interest is withdrawn each year, or the interest is paid automatically, so that the amount invested is always the same.

**EXAMPLE 1**

£3700 is invested at a simple interest rate of 6% for 5 years. Calculate the amount of interest earned.

$$\text{One year's interest} = \frac{6}{100} \times £3700$$
$$= £222$$
$$\text{Five years' interest} = £222 \times 5$$
$$= £1110$$

Simple interest can be calculated using the formula:

$$\textbf{Simple interest} = \frac{\textbf{Rate}}{\textbf{100}} \times \textbf{Principal} \times \textbf{Time}$$

$$I = \frac{R \times P \times T}{100}$$

The formula makes it easier to calculate the rate, the time or the principal.

**EXAMPLE 2**

Calculate the length of time taken for £2000 to earn £270 if invested at 9% simple interest.

The simple interest formula is $I = \dfrac{R \times P \times T}{100}$

Substituting into this gives $270 = \dfrac{9 \times 2000 \times T}{100} = 180 \times T$

$\therefore$ $180 \times T = 270$

Dividing by 180 gives $T = \dfrac{270}{180} = 1.5$ years

**EXERCISE 10.1**

**1** Calculate:

**a** the simple interest on £1400 for 3 years at 6% per annum

**b** the simple interest on £500 at 8.25% per annum for 2 years

**c** the length of time for £5000 to earn £1000 if invested at 10% simple interest per annum

**d** the length of time for £400 to earn £160 if invested at 8% simple interest per annum.

**2** Mr Foyle invests £6750 at 8.5% simple interest per annum.
How much interest has he earned and what is the amount in his account after 4 years?

**3** Mr and Mrs Mahon invest £5000 at 9.25% simple interest per annum for 6 months.
How much interest will they receive?

**4** Mr Allbright invested £10 800 and at the end of each year he withdrew the interest. After 4 years he had withdrawn a total of £3240 in interest. At what annual rate of interest was his money invested?

## 10.2 Compound interest

### Calculation of compound interest

If the interest earned is not withdrawn each year, but is left in the savings account, the amount invested increases as the interest is added. This means that the interest earned the following year will also increase. In five years, with an interest rate of 10% per annum, £1000 'grows' to £1610.51:

£1000 +10% → £1100 +10% → £1210 +10% → £1331 +10% → £1464.10 +10% → £1610.51

This system of paying interest is called **compound interest**. Building societies usually add compound interest to their accounts every year or every six months.

With some bank accounts, e.g. interest-paying current accounts, the interest is calculated daily and added to the account each month.

**EXAMPLE 1**

£3000 is invested in an account paying 12% compound interest per year.

Find the value of the investment after 3 years.

| | |
|---|---|
| Principal | £3000 |
| Interest at 12% (of £3000) | 360 |
| Value of the investment after 1 year | 3360 |
| Interest at 12% (of £3360) | 403.20 |
| Value of the investment after 2 years | 3763.20 |
| Interest at 12% (of £3763.20) | 451.58 |
| Value of the investment after 3 years | £4214.78 |

…sted in a high interest account for one and a half years. The …ate is 10.5% per annum, and it is paid into the account every six …nths.

Calculate the value of the investment after this time and the amount of interest earned.

The rate of interest for 1 year = 10.5%
$\therefore$ The rate of interest for 6 months = 5.25%

| | |
|---|---|
| Principal | £2000 |
| Interest at 5.25% (of £2000) | 105 |
| Value of the investment after 6 months | 2105 |
| Interest at 5.25% (of £2105) | 110.51 |
| Value of the investment after 1 year | 2215.51 |
| Interest at 5.25% (of £2215.51) | 116.31 |
| Value of the investment after $1\frac{1}{2}$ years | £2331.82 |

Compound interest earned = £2331.82 – £2000
= £331.82

Note that at each stage the amount of interest has been rounded to the nearest penny.

## EXERCISE 10.2

1 £6000 is invested at 10% per annum compound interest which is paid annually.
How much is in the account after 3 years?

2 £2000 is invested at 10% per annum payable every 6 months.
How much is in the account at the end of $1\frac{1}{2}$ years?

3 Mrs Fletcher invests £6520 at 9.6% per annum for 18 months. The interest is added every 6 months.
Calculate the total amount of money in the account at the end of 18 months and the amount of interest accrued.

4 £12 600 is invested at 11.05% per annum for 3 years.
Calculate the interest earned over this period.
What is the average amount of interest earned per year?

5 A building society offers a rate of 6.8% per annum payable half-yearly.

a Calculate the interest payable on £1000 at the end of 1 year.

b What is the equivalent yearly interest rate?

6 One building society offers an interest rate of 8.5% per annum payable half-yearly. A second offers 8.75% payable annually.
Mr Dugan has £2000 to invest. Which savings account should he choose?

What other factors (apart from higher interest rate) should Mr Dugan take into account when making his choice?

7 Sharon and Tim put £800 into a high interest account which pays compound interest at the rate of 1.1% per month. After it has gained three months' interest, they withdraw all their money.
How much do they receive? (SEG S96)

## 10.3 Annual Equivalent Rate (AER)

In the example on page 54, we saw that although the interest rate is 10.5% per annum, the amount of interest you would receive if you left your money in the account for a year would in fact be more than 10.5% if interest was calculated every six months.

Having invested £2000, the interest gained after 12 months was £215.51 (with interest calculated every six months).

As a percentage of the money invested, the interest is $\frac{215.51}{2000} \times 100$

$$= 10.775\%$$

This percentage is known as the **Annual Equivalent Rate** or **AER**.

It is defined as the percentage interest gained, as if you only received an interest payment once at the end of the year.

$$\text{Annual Equivalent Rate (AER)} = \frac{\text{Actual Interest gained in one year}}{\text{Amount Invested}} \times 100$$

Notice that when the interest is calculated at more frequent intervals the AER becomes greater.

### EXERCISE 10.3

In Questions 1–6 calculate: **a** the interest earned in 1 year, **b** the AER of the investment.

**1** £4000 is invested at 5.2% per annum. The interest is paid every six months.

**2** £3000 is invested at 5.8% per annum. The interest is paid every six months.

**3** £5000 is invested at 6.3% per annum. The interest is paid every four months.

**4** £8000 is invested at 4.9% per annum. The interest is paid every three months.

**5** £2500 is invested at 6.2% per annum. The interest is paid every three months.

**6** £4000 is invested at 6% per annum. The interest is paid every month.

# 11 Banking

## 11.1 Cheques

All bank current accounts, and some building society accounts, come with a cheque book.

A **cheque** is a written instruction to the bank to pay a sum of money from your account to yourself or to another named person, company or organisation.

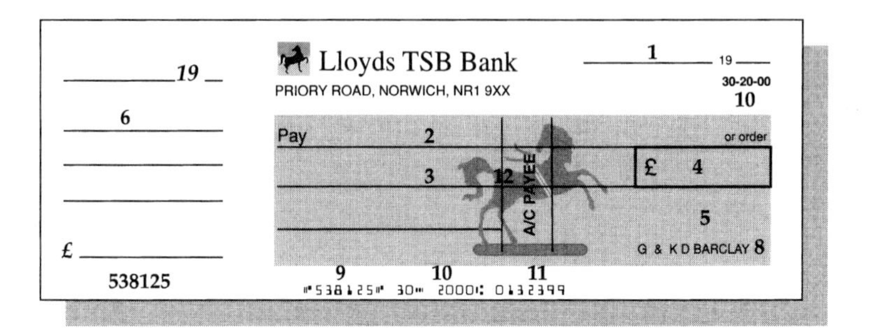


When writing a cheque you must use a pen (**never** a pencil) and fill in correctly:

(1) the date
(2) the name of the person, company or organisation to whom you are paying the cheque (the payee)
(3) the amount of money to be paid, in words (the amount of pence can be written in figures), drawing a line through any unused space
(4) the amount of money to be paid, in figures, separating the pounds from the pence with a hyphen
(5) your signature.

The cheque should be completed neatly and any alterations should be initialled.

You should also fill in:

(6) the counterfoil.

The other information on the cheque is to enable the cheque to be cleared and money transferred from and paid into the correct accounts.

The sorting is done by computers at the Clearing Bank.

The additional information is:

(7) the name and address of the branch of the bank that holds the account
(8) the name on the account
(9) the number of the cheque, which will appear on the bank statement
(10) the branch's sorting code number
(11) the account number
(12) the two lines crossing the cheque vertically – all cheques issued by the banks are 'crossed', which means the cheque must be paid into a bank account.

**EXERCISE 11.1**

1 The cheque shown below was written by R S Langdon to M F Townsend for the amount of £43, on 7 March 1999.

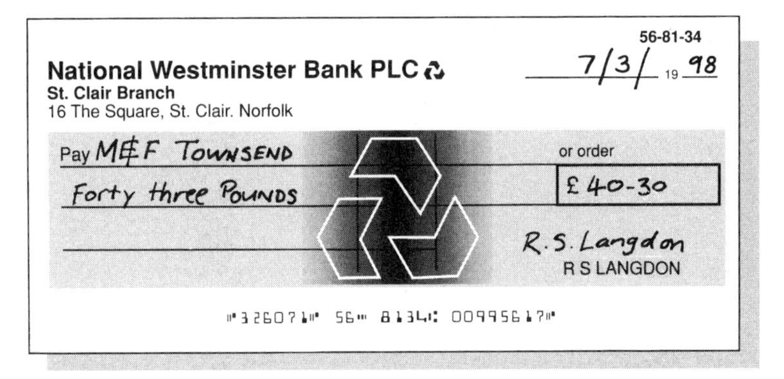

56-81-34

National Westminster Bank PLC
St. Clair Branch
16 The Square, St. Clair. Norfolk

7/3/ 19 98

Pay M&F TOWNSEND or order

Forty three POUNDS

£ 40-30

R.S. Langdon
R S LANGDON

Unfortunately, several mistakes have been made. For each mistake you find, explain what is wrong and what should have been written.

2 Give four reasons why a bank might return a cheque unpaid.

3 What do the words 'refer to drawer' mean, when written on a cheque?

4 Explain the meaning of the phrases 'A/C payee' and 'not negotiable' when written on a cheque.

5 What benefit is there in possessing a cheque guarantee card?

6 Why should a cheque book and a cheque card always be kept separate?

## 11.2 Current accounts

For most people, a current account is the main bank account. It is a convenient way of dealing with everyday income and expenditure because the bank offers many useful services in conjunction with a current account. Here are some of them:

- cheque book
- cheque guarantee card
- standing orders
- direct debits
- cash dispenser card
- payment debit card
- monthly statement

Standing orders and direct debits are usually made from a bank account, normally a current account, but the service is also available with some savings accounts.

## Standing orders

The account holder instructs the bank to make regular payments, of a specified amount, on his/her behalf.

**Standing orders** can be used to pay insurance premiums, hire purchase instalments or to transfer regular amounts of money to a savings or investment account.

Because the amount to be paid can only be changed by the **payer**, standing orders are being superseded by the more flexible **direct debit**.

## Direct debits

Direct debits can be used to pay regular, fixed amounts in the same way as standing orders, but, because the **payee** can change the amount to be paid, variable payments can be made at variable intervals.

The account holder (payer) is usually given 14 days' notice of any change in the amount to be paid.

Direct debits are used to pay subscriptions or bills such as gas, electricity, telephone, Council Tax, where the amounts are likely to change each year. The payment of a direct debit is at the request of the **payee's bank** and *not* the account holder or the account holder's bank.

## Cash dispenser card

As an account holder at a bank or building society, you can apply for a **cash dispenser card**. You will also be given a PIN (personal identity number) which, together with the card, will enable you to obtain cash from a cashpoint machine (also called an ATM, automated teller machine).

Most of these machines dispense money outside banking hours and many operate during the night.

## Payment debit card

A debit card can be used to buy goods from a shop, and it works in a similar way to a cheque. The money is deducted directly from your bank account within a few days.

It can also be used to obtain money from cash dispenser machines or over the counter at a bank.

## Bank statements

The bank will send an account holder a **bank statement** each month. This is a concise record of all payments into (credits) and out of (debits) his or her account during the preceding month.

The statement should be checked carefully.

## EXERCISE 11.2

**1** Study the bank statement and then answer the following questions:

**a** With which branch does the account holder bank her money?

**b** What is the account number?

**c** On what date was the statement produced?

**d** What was the balance in the account on 13 March?

**e** By how much did the account become overdrawn?

**f** How much was withdrawn using cashpoint machines?

**g** What is the meaning of D/D?

**h** How much was paid on standing orders?

**i** What does the symbol * mean?

**j** What are the numbers 412– – –?

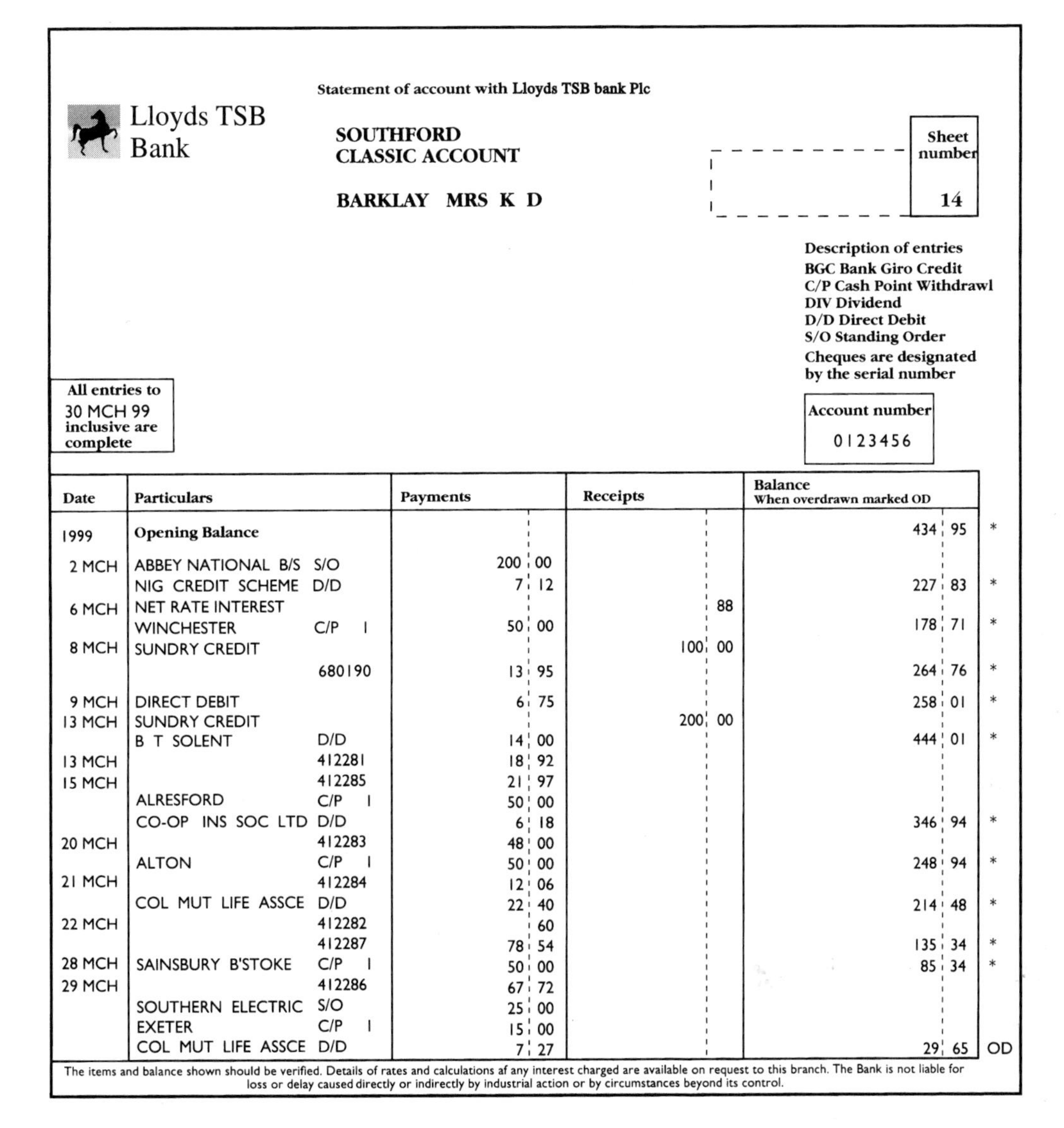

Statement of account with Lloyds TSB bank Plc

Lloyds TSB Bank

SOUTHFORD
CLASSIC ACCOUNT

BARKLAY MRS K D

Sheet number 14

Description of entries
BGC Bank Giro Credit
C/P Cash Point Withdrawl
DIV Dividend
D/D Direct Debit
S/O Standing Order
Cheques are designated by the serial number

All entries to 30 MCH 99 inclusive are complete

Account number 0123456

| Date | Particulars | | Payments | Receipts | Balance When overdrawn marked OD | |
|---|---|---|---|---|---|---|
| 1999 | Opening Balance | | | | 434 95 | * |
| 2 MCH | ABBEY NATIONAL B/S | S/O | 200 00 | | | |
| | NIG CREDIT SCHEME | D/D | 7 12 | | 227 83 | * |
| 6 MCH | NET RATE INTEREST | | | 88 | | |
| | WINCHESTER | C/P I | 50 00 | | 178 71 | * |
| 8 MCH | SUNDRY CREDIT | | | 100 00 | | |
| | | 680190 | 13 95 | | 264 76 | * |
| 9 MCH | DIRECT DEBIT | | 6 75 | | 258 01 | * |
| 13 MCH | SUNDRY CREDIT | | | 200 00 | | |
| | B T SOLENT | D/D | 14 00 | | 444 01 | * |
| 13 MCH | | 412281 | 18 92 | | | |
| 15 MCH | | 412285 | 21 97 | | | |
| | ALRESFORD | C/P I | 50 00 | | | |
| | CO-OP INS SOC LTD | D/D | 6 18 | | 346 94 | * |
| 20 MCH | | 412283 | 48 00 | | | |
| | ALTON | C/P I | 50 00 | | 248 94 | * |
| 21 MCH | | 412284 | 12 06 | | | |
| | COL MUT LIFE ASSCE | D/D | 22 40 | | 214 48 | * |
| 22 MCH | | 412282 | 60 | | | |
| | | 412287 | 78 54 | | 135 34 | * |
| 28 MCH | SAINSBURY B'STOKE | C/P I | 50 00 | | 85 34 | * |
| 29 MCH | | 412286 | 67 72 | | | |
| | SOUTHERN ELECTRIC | S/O | 25 00 | | | |
| | EXETER | C/P I | 15 00 | | | |
| | COL MUT LIFE ASSCE | D/D | 7 27 | | 29 65 | OD |

The items and balance shown should be verified. Details of rates and calculations af any interest charged are available on request to this branch. The Bank is not liable for loss or delay caused directly or indirectly by industrial action or by circumstances beyond its control.

**2** Explain briefly how you would check the above bank statement.

**3** Name six services or facilities available when you open a bank account.

# 12 Borrowing and Spending

## 12.1 True interest rates

Someone who wants to spend money may sometimes need to borrow money or buy on credit.

The cost of borrowing can vary a great deal depending on the method of obtaining credit as well as the amount borrowed.

Usually banks, stores and finance companies quote a flat rate of interest, but this is *not* the true price of the credit.

So that customers can accurately compare credit costs, the **APR** (**annual percentage rate**) must, by law, be quoted for all credit arrangements.

This is the *true* rate of interest, and it takes into account all the costs involved and the method of repayment. (In general, the lower the APR, the better the deal.)

If you borrow £100 for one year and the 'credit charge' is 8% you repay £108 (the loan plus interest).

But the charge of £8 would only be a true rate of 8% if you had the use of the £100 for the whole year. In fact, you have £100 at the start of the year and nothing at the end.

In practice, you begin repaying the loan almost immediately in twelve equal instalments spread over the year.

The average amount available to you during the year is therefore about half the original amount and the **true rate of interest is nearly double the quoted rate**:

$$\text{APR} \approx 2 \times \text{Flat rate of interest}$$

The APR is, in fact, between 1.8 and 2 times the flat rate of interest as the following example shows:

Suppose you borrow £100 for 1 year at a flat rate of interest of 8% per annum.

The total repayment = £100 + £8 interest = £108.

This is repaid in 12 monthly instalments of £9 (= £108 ÷ 12).

If the interest on the outstanding debt is calculated at a rate of 1.205% per month, the debt will be wiped out by the end of the year, as shown on the following page:

| | £ |
|---|---|
| Initial debt | 100.00 |
| Interest in first month at 1.205% per month | 1.21 |
| Debt at end of first month | 101.21 |
| £9 repaid. Debt at beginning of second month | 92.21 |
| Interest in second month on a debt of £92.21 | 1.11 |
| Debt at end of second month | 93.32 |
| £9 repaid. Debt at beginning of third month | 84.32 |
| Interest in third month | 1.02 |
| Debt at end of third month | 85.34 |
| £9 repaid. Debt at beginning of fourth month | 76.34 |
| Interest in fourth month | 0.92 |
| Debt at end of fourth month | 77.26 |
| £9 repaid. Debt at beginning of fifth month | 68.26 |
| Interest in fifth month | 0.82 |
| Debt at end of fifth month | 69.08 |
| £9 repaid. Debt at beginning of sixth month | 60.08 |
| Interest in sixth month | 0.72 |
| Debt at end of sixth month | 60.80 |
| £9 repaid. Debt at beginning of seventh month | 51.80 |
| Interest in seventh month | 0.62 |
| Debt at end of seventh month | 52.42 |
| £9 repaid. Debt at beginning of eighth month | 43.42 |
| Interest in eighth month | 0.52 |
| Debt at end of eighth month | 43.94 |
| £9 repaid. Debt at beginning of ninth month | 34.94 |
| Interest in ninth month | 0.42 |
| Debt at end of ninth month | 35.36 |
| £9 repaid. Debt at beginning of tenth month | 26.36 |
| Interest in tenth month | 0.32 |
| Debt at end of tenth month | 26.68 |
| £9 repaid. Debt at beginning of eleventh month | 17.68 |
| Interest in eleventh month | 0.21 |
| Debt at end of eleventh month | 17.89 |
| £9 repaid. Debt at beginning of twelfth month | 8.89 |
| Interest in twelfth month | 0.11 |
| Debt at end of twelfth month | 9.00 |
| £9 repaid. Debt after twelve repayments of £9 | 0.00 |

The interest rate of 1.205% per month gives a compound interest rate of 15.46% per year.

Hence a flat rate of 8% per year is equivalent to an APR of 15.46% (almost twice the percentage rate).

(How did we get that figure of 1.205% per month? That involves some tricky arithmetic which you needn't worry about!)

## 12.2 Buying on credit

If you do not have sufficient funds saved to buy the goods you require, or do not wish to withdraw a large sum of money from your savings, you can buy the goods on credit, i.e. borrow money which will be repaid over a period of time, usually in instalments which include interest.

There are several methods of obtaining credit:

- Hire purchase
- Credit purchase
- Loans
- Monthly accounts
- Credit card/Store card/Charge card.

## Hire purchase and credit purchase

These two methods of obtaining credit are similar. A **deposit** is often required and the amount borrowed plus interest is repaid in equal instalments over a set period of time.

The main difference is that with hire purchase you do not own the goods until the last payment has been made and so cannot sell the goods while you are still paying for them. Also, the hire purchase company can repossess the goods if the repayments are not made.

Credit purchase is usually arranged with the store from which you buy the goods. You own the goods from the moment the agreement is signed. If you fail to keep up the repayments, the store will sue through the courts for the amount outstanding.

### EXAMPLE 1

A video recorder can be bought for a cash price of £299.95 or by credit purchase paying 12 monthly instalments of £31.50.
Calculate:

**a** the credit price

**b** the interest paid

**c** the flat rate of interest, as a percentage of the amount borrowed

**d** the approximate APR.

**a** Credit price = Total instalments
= 12 × £31.50
= £378

**b** Interest paid = Credit price − Cash price
= £378.00 − £299.95
= £78.05

**c** $$\text{Flat rate of interest} = \frac{\text{Interest}}{\text{Amount borrowed}} \times 100\%$$

$$= \frac{£78.05}{£299.95} \times 100\%$$

$$= 26.0\%$$

**d** Approximate APR = 2 × Flat rate per year
= 52.0%

**EXAMPLE 2**

A washing machine is offered for a cash price of £329.99.
It can also be bought on hire purchase for a deposit of £34.99 and 30 monthly instalments of £13.25.

Calculate:

**a** the total hire purchase price
**b** the amount of interest paid
**c** the amount borrowed
**d** the flat rate of interest
**e** the approximate APR.

**a** Total HP price = Total instalments + Deposit (do not forget to include the deposit)
= 30 × £13.25 + £34.99
= £397.50 + £34.99
= £432.49

**b** Interest paid = Credit price – Cash price
= £432.49 – £329.99
= £102.50

**c** Amount borrowed = Cash price – Deposit
= £329.99 – £34.99
= £295.00

**d** $\text{Flat interest rate} = \dfrac{\text{Interest}}{\text{Amount borrowed}} \times 100\%$

$= \dfrac{£102.50}{£295} \times 100\%$

$= 34.7\%$

**e** $\text{Flat rate per year} = \dfrac{\text{Flat rate}}{\text{Time of loan in years}}$

$= \dfrac{34.7\%}{2.5}$

$= 13.9\%$

Approximate APR = 2 × Flat rate per year
= 2 × 13.9%
= 27.8%

## EXERCISE 12.1

**1** A department store offers a hi-fi system for sale at £429.99. The customer can also buy the system on interest-free credit by paying a deposit of 10% of the cash price and 10 equal monthly instalments.

**a** How much is the deposit?

**b** How much is each monthly payment?

**2** A rival store offers a similar hi-fi system for sale at £349.99 or 12 months' free credit. The credit terms are 10% deposit and 12 equal monthly instalments.

How much is each instalment?

**3** A television set can be bought for £189.99 cash or by credit purchase.

The credit purchase terms are: no deposit and 24 monthly payments of £9.31. Calculate:

**a** the total credit price
**b** the amount of interest paid
**c** the flat rate of interest
**d** the approximate APR.

4 A motorbike has a cash price of £4195. It can be bought on hire purchase for a deposit of 10% and 36 monthly payments of £135.84.
Calculate:

**a** the total hire purchase cost

**b** the approximate APR.

5 Mr Weston is buying a new car. The cash price of the car is £7399. He is offered a £1500 trade-in price on his old car. The hire purchase terms on the new car are a deposit of 25% of the cash price and 24 monthly payments of £290.
Calculate:

**a** the deposit

**b** the amount borrowed by Mr Weston from the HP company

**c** the total hire purchase cost of the car

**d** the interest paid, as a percentage of the loan

**e** the approximate APR.

6 Mrs Sayeed wishes to buy a new car. The cash price of the car is £9098.

The hire purchase terms offered by the garage are: a deposit of 15% and *either* 36 monthly repayments of £271.50 *or* 60 monthly repayments of £179.12.

Find, for each deal:

**a** the total hire purchase price of the car

**b** the approximate APR.

7 The same camera is available in two different shops at the same price of £98.60, but with different credit terms.

Shop A requires a deposit of 10% and twelve monthly repayments of £8.43.

Shop B requires a deposit of 15% and ten monthly repayments of £9.60.

By calculating the true rates of interest (APR), decide which shop is offering the better deal.

8 A car is offered for sale in a garage with a price of £8400 on the windscreen. John can either pay by cash or credit.
If John buys the car on credit, he pays a deposit of 20% and 36 monthly instalments of £210.

**a** (i) Find the deposit he would have to pay.
(ii) Find the total credit price.

If John pays cash he will pay the *cash price* which will give John a discount of $12\frac{1}{2}$% off the windscreen price.

**b** What is the extra amount paid for credit compared with cash?

**c** If John buys the car on credit he requires a loan. The loan will be the difference between the cash price and the deposit.
(i) Express the extra amount paid for credit compared with cash as a percentage of the loan.
(ii) Hence make an estimate of the APR which John is charged for credit.

(SEG Spec98)

## 12.3 Interpolation

If you wish to take out a loan, you will problably be given a table which shows the monthly repayments required if the load is taken out for different periods of time, such as 12 months, 24 months or 36 months.

Using the method of interpolation described below, it is possible to obtain an estimate for the monthly repayments needed to pay back a loan if it it taken out for different periods of time from those shown on the table.

### EXAMPLE

The monthly repayment on a loan of £6000 repayable over 24 months is £306.40 and over 36 months is £220.93. By interpolation, estimate the repayment if the loan is taken out for 32 months.

By plotting these values on a graph, we can see that the monthly repayment is reduced by £306.40 – £220.93 = £85.47 for repaying the loan over 36 months instead of 24, (i.e. repaying for an extra 12 months).

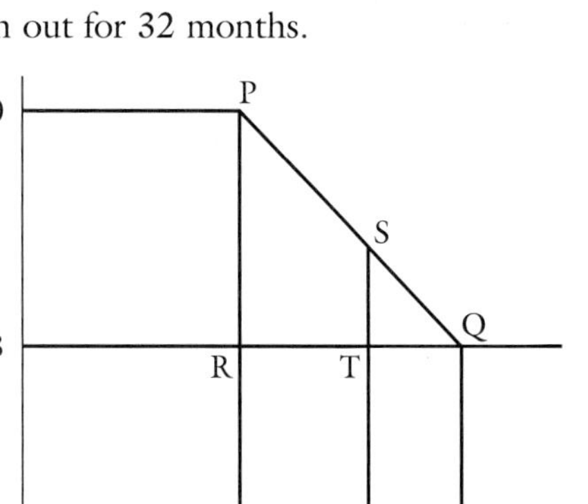


By drawing a straight line, joining the monthly repayments after 24 months and 36 months, we can estimte the repayment after any number of months. In the diagram, the point *S* identifies the monthly repayment for 32 months.

Drawing the line RQ gives two similar triangles STQ and PRQ

Thus $\frac{ST}{TQ} = \frac{PR}{RQ}$

$$ST = \frac{PR}{RQ} \times TQ$$

$$= £\frac{85.47}{12} \times 4$$

$$= £28.49$$

Thus the repayment over a 32 month period is the repayment at Q + ST

$$= £220.93 + £28.49$$

$$= £249.42$$

## EXERCISE 12.2

In Questions 1–6, use the table on monthly repayments given below to estimate by interpolation:

1 the monthly repayment on a loan of £4000 for a period of 18 months

2 the monthly repayment on a loan of £5000 for a period of 30 months

3 the monthly repayment on a loan of £3000 for a period of 20 months

4 the monthly repayment on a loan of £6000 for a period of 44 months

5 the monthly repayment on a loan of £10 000 for a period of 28 months

6 the monthly repayment on a loan of £5500 for a period of 16 months

| Period of loan | Amount of loan: | £1000 | £2000 | £3000 | £4000 | £5000 | £6000 |
|---|---|---|---|---|---|---|---|
| 12 months | | 91.01 | 182.01 | 270.56 | 360.74 | 450.93 | 535.14 |
| 24 months | | 49.24 | 98.48 | 145.24 | 193.66 | 242.07 | 284.49 |
| 36 months | | 35.44 | 70.89 | 103.77 | 172.95 | 172.95 | 201.39 |
| 48 months | | 28.64 | 57.28 | 83.26 | 138.78 | 138.78 | 160.16 |

## 12.4 Discrimination in spending

It pays to be discriminating when spending money. Taking time to make comparisons means getting the best value for your money.

You can save money by looking for special offers, buying during the sales (if you are sure you are getting a bargain) or buying in bulk.

### EXAMPLE

Mrs Newman is shopping for soap powder and finds a new brand on the supermarket shelves.

The powder is sold in three packet sizes:

a starter pack: weight 145 g, price 39p
a medium size: weight 800 g, price £1.55
a large size: weight 2 kg, price £3.79

Which size should Mrs Newman buy to give her the best value for her money?

There are two methods which can be used to compare value for money on the basis of cost.

*Method 1*

Prices are compared by calculating the cost of 1 g of powder for each size (given here to 3 significant figures).

1 g costs:

(i) Starter pack $\frac{39\text{p}}{145} = 0.269\text{p}$

(ii) Medium size $\frac{£1.55}{800} = 0.194\text{p}$

(iii) Large size $\frac{£3.79}{2000} = 0.190\text{p}$

| Packet type: | *Starter* | *Medium* | *Large* |
|---|---|---|---|
| Price per g: | 0.269p | 0.194p | 0.190p |

This shows that the large size gives the best value for price.

*Method 2*

Value is compared by calculating the amount of powder which can be bought for the same amount of money (e.g. £1) in each case (again to 3 significant figures).

Amount of powder for £1:

(i) Starter pack $\frac{145}{0.39} = 372\text{g}$

(ii) Medium size $\frac{800}{1.55} = 516\text{g}$

(iii) Large size $\frac{2000}{3.79} = 528\text{g}$

| Packet type: | *Starter* | *Medium* | *Large* |
|---|---|---|---|
| Price per £1: | 372 g | 516 g | 528 g |

This also shows that the large size will give Mrs Newman more for her money.

Of course, price is not the only thing to be taken into account when making a purchase. It is also important to consider, where appropriate:

- the quality of the product
- the length of time for which non-durables can be stored
- personal preferences of taste, colour, style etc.
- the manufacturer's reputation for reliability
- the shop's reputation for service.

For example, Mrs Newman might consider that the starter pack will be the best value at the moment. If she finds that she prefers her old powder, she has not spent much money on a product she does not want.

The medium size packet costs very little more per kg than the large size and is lighter to carry. It also takes up less storage space. For some people, this size may be the best buy.

## EXERCISE 12.3

**1** **a** A well-known brand of margarine sells at 38p for 250 g and 67p for 500 g.

(i) For how much per kg is each size sold?
(ii) Which size is the better value for money?

**b** The price of the smaller size of margarine is reduced by 5p on 'special offer'. Which size is the best buy?

**2** Mayonnaise costs 99p for a 400 g jar and £1.39 for a 600 g jar. Mrs Kaplan has a voucher for '10p OFF' a jar of mayonnaise.

**a** If she uses the voucher, how many grams of mayonnaise will cost 1p:

(i) with the 400 g jar
(ii) with the 600 g jar?

**b** Which jar gives the best value for money?

**c** What other factor(s) should Mrs Kaplan take into consideration before making her purchase?

**3** The village shop stocks soap powder in three sizes:

E3 contains 1.05 kg of powder and costs £1.22
E10 contains 3.50 kg of powder and costs £3.79
E15 contains 5.25 kg of powder and costs £5.49.

**a** Calculate comparative costs for each size, based on the E15 size.

**b** Which size would you recommend as the best buy for the following people, giving at least one reason for each choice?

(i) Mr and Mrs Donovan and their three children
(ii) Mrs Kyle, a 66-year-old widow
(iii) Mr and Mrs Groves, a newly married couple with no children, but a large mortgage
(iv) Mr Hambledon, a bachelor living in a small flat.

**4** Compare the following advertisements and decide which is the best value for money, giving clear reasons for your choice.

# 13 Household Bills

Substantial bills which a householder can expect to receive each year are:

- Water rates
- Electricity bill
- Gas bill
- Telephone bill

The electricity, gas and telephone bills arrive quarterly; the water rates bill arrives annually.

## 13.1 Electricity bills

The amount of electricity or gas which a household uses is recorded by a meter.

There are two types of meter in use.

The digital meter is a line of figures:

The clock meter has dials which are read from left to right. Each dial is numbered from 0 to 9 and the hands move clockwise and anticlockwise alternately.

When a hand is between numbers the lower number is recorded. (When a hand is between 9 and 0, record 9.)

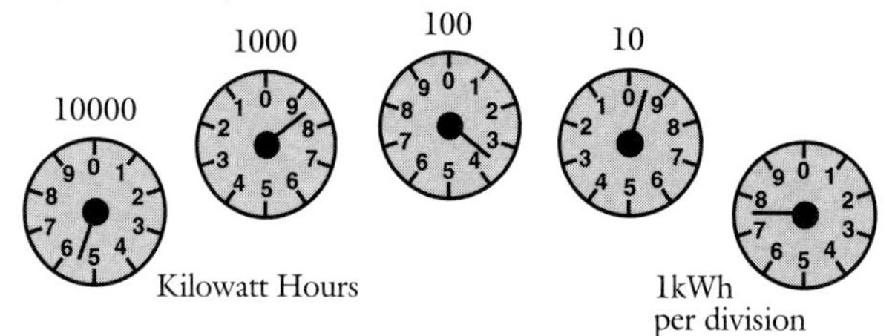


To calculate an electricity bill the price per unit is multiplied by the number of units used and a standing charge is added. VAT is charged at 5% on electricity.

**EXAMPLE 1**

What is the reading on the set of dials above?

The reading is 58397.

**EXAMPLE 2**

In February, the reading on Mr Spencer's meter is 28279 and the previous reading, in November, was 26814.

The cost of electricity is 5.44p per unit and the quarterly standing charge is £6.27.

Calculate:

**a** the number of units of electricity used

**b** the cost of electricity used

**c** the total electricity bill inclusive of 8% VAT.

a Number of units used = 28 279 − 26 814 = 1465

b Cost of electricity = 1465 × 5.44p = £79.70

c Total bill (excluding VAT) = £79.70 + £6.27 = £85.97

VAT at 5% = £4.30

Total bill (including VAT) = £90.27

## EXERCISE 13.1

**1** What is the reading on the following sets of dials?

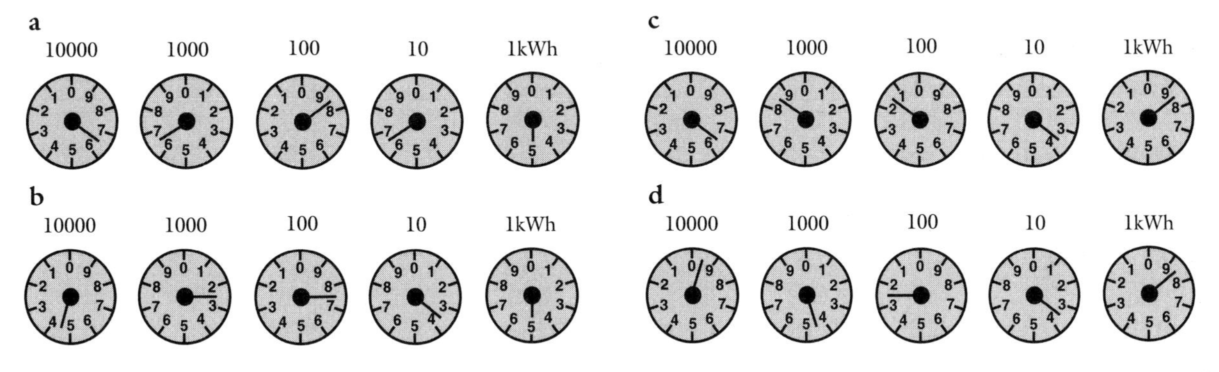


**2** Calculate the total electricity bills, including VAT at 5%, for the following amounts:

| | Present reading | Previous reading | Cost per unit | Standing charge |
|---|---|---|---|---|
| a | 37162 | 35841 | 5.90p | £9.70 |
| b | 86719 | 85017 | 6.83p | £9.90 |
| c | 26341 | 24863 | 5.70p | £7.67 |
| d | 70905 | 69438 | 6.04p | £8.45 |
| e | 42186 | 41902 | 7.63p | £9.20 |

In questions **3** and **4**, assume all prices given are inclusive of VAT.

**3** Mr Sinclair receives the following electricity bill based on an estimated reading:

| Previous reading | Present reading | Tariff | Units | Price per unit | Amount £ | p |
|---|---|---|---|---|---|---|
| 77945 | 78616E | DOMESTIC | 671 | 6.04p | 40 | 53 |
| QUARTERLY CHARGE | | | | | 8 | 45 |
| | | TOTAL THIS ACCOUNT | | | 48 | 98 |

The true reading was 78262.

a By how much has Mr Sinclair been overcharged?

b Draw a set of dials and put in arrows to show the true meter reading.

**4** Mrs Yamoto receives an electricity bill for £59.08.
The standing charge is £7.67 and the cost per unit is 5.70p.

a How many units of electricity were used?

b If the present reading is 66112, what was the previous reading?

## 13.2 Gas bills

Gas bills are calculated in a similar way to electricity bills, but the number of units used is the number of cubic feet of gas used.

This number has then to be converted into therms and multiplied by the price per therm.

A standing charge is added.

Typically 105 therms = 100 cubic feet of gas.

**EXAMPLE**

In one quarter Mr and Mrs Blake used 260 hundred cubic feet of gas. The cost of the gas was 36p per therm and the standing charge was £8.98 per quarter.

Calculate the amount paid for gas.

Amount of gas used = 260 hundred cubic feet of gas

$= \frac{260}{100} \times 105$ terms

= 273 therms

Cost of gas = 36p × 273

= £98.28

Standing charge = £8.98

Total = £107.26

VAT at 5% = £5.36

Amount paid (incl. VAT) = £112.62

*EXERCISE 13.2*

1 In the following questions, the cost of gas is 39.8p per therm and the standing charge is £8.70 per quarter.

Calculate the quarterly gas bills (excluding VAT) for the following people:

**a** Mr Hudson, who used 401 therms
**b** Mrs Snell, who used 275 therms
**c** Miss Dawe, who used 392 therms
**d** Mr Ghosh, who used 456 therms.

2 Mrs Kirk used 370 hundred cubic feet of gas in one quarter. How many therms did she use?

**3** Mr McBain used 95 therms of gas at a cost of 34.2p per therm. The standing charge was £9.00 per quarter.
How much was his gas bill (including VAT)?

**4** Miss Hennigan's gas meter reads 3469 on 10 December and 3617 on 10 March. The charge per therm is 43p and the quarterly standing charge is £9.40.

Calculate:

**a** the number of units of gas consumed

**b** the number of therms of gas consumed

**c** the cost of the gas

**d** the total gas bill for the quarter (including VAT).

**5** **a** The reading on Stephen Bond's gas meter at the beginning of the quarter was 5628.

At the end of the quarter the reading is as shown below:

(i) What reading is shown on the meter?
(ii) How many units (i.e. hundreds of cubic feet of gas) have been used during the quarter?

**b** The cost of gas is 39.8p per therm and there is a standing charge of £8.70 per quarter.
(i) How many therms have been consumed?
(ii) What is the cost of the gas consumed?
(iii) What is the total gas bill (including VAT)?

## 13.3 Telephone bills

Telephone bills are sent three or four times per year. You are charged for the use of the system, the hire of the telephone, and the calls which have been made.

Some people buy their own telephone, and their bills will not include the charge for the hire of the telephone.

VAT is added to telephone bills as a service charge.

## EXERCISE 13.3

1 Mrs Kingsley has her own telephone. She pays the system charge of £21.86 and the cost of 415 units at 4.20p per unit. Calculate her total bill (excluding VAT).

2 A meter reading on 1 March was 27258, and on 31 May it was 27623. The cost per unit was 4.20p and the standing charges totalled £21.86. Calculate the total bill (excluding VAT).

3 a Mr Stears' telephone bill is made up of the following charges:
- quarterly rental, £21.86
- 196 metered units @4.20p per unit
- operator calls, £1.68

Calculate the total bill (excluding VAT).

b VAT at 17.5% is added to the bill. Calculate the final bill which Mr Stears receives.

4 Complete the following telephone bill:

| Quarterly rate | | | £ |
|---|---|---|---|
| SYSTEM | | | 21.86 |
| APPARATUS | | | 3.10 |
| TOTAL | | | a |
| Date | Meter reading | Units used | |
| 13 MAR | 024367 | | |
| 12 JUN | 024536 | b | |
| UNITS AT 4.20p | | | c |
| TOTAL (EXCLUSIVE OF VAT) | | | d |
| VAT AT 17.5% | | | e |
| TOTAL PAYABLE | | | f |

# 14 Further Percentages

## 14.1 Inverse percentages

Inverse percentages were used in Unit 8 to find the original price of an item when the price inclusive of VAT was known. In this unit the same method is used in a more general context.

**EXAMPLE**

Andy bought a guitar which he later sold to a friend for £57.40, making a loss of 18% on the amount he paid.

How much did the guitar originally cost?

The **original price** is always represented by **100%**

Andy's selling price is represented by $100\% - 18\% = 82\%$

So 82% represents the selling price of £57.40

1% represents $\frac{£57.40}{82}$

100% represents the cost price of $\frac{£57.40}{82} \times 100 = £70$

Therefore the guitar originally cost £70.

If preferred, the information can be represented on a table instead:

| | % | £ |
|---|---|---|
| Cost price | 100 | ? |
| Loss | 18 | ? |
| Selling price | 82 | 57.40 |

$$\text{Cost price} = \frac{£57.40}{82} \times 100 = £70$$

If required, the amount of money lost on the deal could be found by a similar method:

$$\text{Loss} = \frac{£57.40}{82} \times 18 = £12.60$$

### EXERCISE 14.1

1 In the following questions you are given the percentage profit and the selling price. For each case find the original cost price.

a 20%, £30
b 28%, £69.12
c $12\frac{1}{2}$%, £1662
d 110%, £466.20
e 28.7%, £57.92

2 In the following questions you are given the percentage loss and the selling price. For each case, find the original cost price.

a 10%, £54
b $7\frac{1}{2}$%, £31.45
c 5%, 96p
d 28%, £864
e 8.25%, £77.99

3 Doris is a pensioner and does not pay income tax. Her bank account earns interest of £540 from which tax at a rate of 25% has been deducted. How much tax can she reclaim?

4 Mario's wage is increased by 9.2%. He now earns £235 per week. How much did he earn previously?

5 After one year Mrs Brennan's car is valued for insurance at £5600, a depreciation on the price when new of 22%. How much, to the nearest £10, did Mrs Brennan pay for the car?

6 A garden centre buys plants which it resells at a profit of 28%. How much was the original price of a rose bush which is sold for £3.40?

7 A car is sold for £5225 after a depreciation of 45% of the original purchase price. Calculate the original purchase price of the car. (SEG Spec 98)

8 In a sale, everything is reduced by 15% of the marked price. Kerry was given a reduction of £7.49 on a cassette player. What was its marked price? (SEG W93)

## 14.2 Compound percentages

**EXAMPLE**

A wholesaler adds 20% profit to the price of goods when he sells to a retailer. The retailer then adds 15% profit to the same goods before selling to the customer. What percentage above the original price does the customer pay?

The answer is NOT 35%!

The original price is equivalent to 100%
The wholesale price is equivalent to $100 + 20\%$ of $100 = 120\%$
The retail price is equivalent to $120 + 15\%$ of $120 = 138\%$

The customer pays 38% above the original price.

**EXERCISE 14.2**

1 The wholesale price of a certain item is 25% more than the manufacturing cost. The retail price is 20% more than the wholesale price.

What is the percentage difference between the retail price and the manufacturing cost?

2 Mr Miles invests a sum of money for 2 years at an interest rate of 10% per annum.

By what percentage has his original sum of money increased after two years?

3 The menswear department is having a sale. The normal selling price of a particular style of shirt includes a profit of 30%. During the sale the normal price is reduced by 15%.

What percentage profit does the shop make on these shirts during the sale?

4 Last year when Mrs Berry had her car serviced, 80% of the cost of servicing was for labour and 20% was for parts. In one year the labour costs increased by 9% and the cost of parts by 12%. What is the percentage increase in Mrs Berry's bill compared to the previous year?

5 The manufacturing cost of a child's toy is made up of 50% labour, 30% materials and 20% overheads. The factory improves the machinery which increases the overheads by 80%, but decreases the labour costs by 25%. At the same time the cost of materials increases by 10%.

What is the overall percentage increase in the manufacturing cost?

6 A motor trader predicts that his s in 1997 and buys 45% more vehi predictions were incorrect and h sell 28% of the total vehicles boı Calculate the percentage increa 1997.

(SEG Spec 98)

## 14.3 Depreciation

Most possessions, such as cars, caravans, electrical equipment, depreciate in value as time passes.

For example, at the end of each year the value of a car will be less than its value at the beginning of the year.

Depreciation is calculated by a similar method to compound interest.

**EXAMPLE**

A car was bought for £6500 in 1990. During the first year of ownership its value depreciated by 20% and during each subsequent year by 15%.

Calculate the value of the car three years later.

| | £ |
|---|---|
| Cost of car | 6500 |
| Depreciation of 20% (of £6500) | 1300 |
| Value after 1 year | 5200 |
| Depreciation of 15% (of £5200) | 780 |
| Value after 2 years | 4420 |
| Depreciation of 15% (of £4420) | 663 |
| Value after 3 years | 3757 |

The value of the car after 3 years is £3757.

*EXERCISE 14.3*

1 A small business buys a computer costing £4500. The rate of depreciation is 20% per annum. What is its value after three years?

2 A motorbike is bought second hand for £795. Its price depreciates by 11% per year. For how much could it be sold two years later? (Give your answer to the nearest pound.)

3 Mr and Mrs Parsons' carpet is accidentally damaged by fire. It was bought only three years ago for £450. The insurance investigator (loss-adjuster) decides that it will have depreciated in value by 10% each year.

What was the value of the carpet just before the accident?

4 Mr Carey invests £3000 in unit trusts in the hope that they will appreciate in value. Unfortunately, although they appreciate in value by 8% during the first year, they depreciate in value during the second year by 8% and during the following two years depreciate by 10% of the value at the beginning of each year.

How much are the unit trusts worth after four years?

## Mark-up

hen a retailer buys goods to sell to customers, the retailer increases its price to cover costs and to make a profit.

This increase, expressed as a percentage of the retailer's cost price, is called the **retailer's mark-up**.

**EXAMPLE**

A retailer paid £80 for a suit and sold it for £126. Find the percentage mark up.

The increase in price is £126 – £80 = £46

The percentage mark-up is $\frac{46}{80} \times 100 = 57.5\%$

**Percentage mark-up** $= \frac{\textbf{Increase in price}}{\textbf{Cost price}} \times \mathbf{100}$

**EXAMPLE**

A shopkeeper has a normal mark-up of 90%. In a sale all prices are reduced by 40%. Find the actual mark-up during the sale.

Taking the shopkeeper's cost price as 100%,
the normal increase in price is 90% of the shopkeeper's cost price,
and hence the selling price is 190% of the shopkeeper's cost price.

Discount in the sale is 40% of the normal selling price,
which is 40% of the usual 190% of the shopkeeper's cost price

$= \frac{40}{100} \times 190\%$ of the shopkeeper's cost price

$= 76\%$ of the shopkeeper's cost price.

$\therefore$ Selling price is $190 - 76\% = 114\%$ of the shopkeeper's cost price

$\therefore$ Actual mark-up is 14%

### *EXERCISE 14.4*

1 A dress is sold at £70. The cost to the retailer was £50. What is the percentage mark-up?

2 A pair of trousers is sold at £65. The cost to the retailer was £40. What is the percentage mark-up?

3 A bookseller has a normal mark-up of 80%. During a sale all prices are reduced by 30%. Find the actual mark-up during the sale.

4 A shoe shop has a normal mark-up of 70%. During a sale all prices are reduced by 25%. Find the actual mark-up on the goods in the sale.

5 A jeweller has a normal mark-up of 90%. Because of increased demand the prices are increased by 20%. Find the actual mark-up after this increase.

6 A sports shop has a normal mark up of 75%. During a sale all prices are reduced by 30%. Find the actual mark-up on the goods in the sale.

# 15 Statistical Terms

In 1834, the Royal Statistical Society was founded, and defined statistics as 'using figures and tabular exhibitions to illustrate the conditions and prospects of society'. Statistics is now used to deal with the collection, classification, tabulation and analysis of information and opinions.

**Data.** Data is the information which has been collected or researched. The word 'data' is a plural and the singular is 'datum' (a single piece of information).

**Variables.** Information is collected about *variables* such as weights, numbers of clients, types of disease.

A *variable* is something which can change from one item to the next. It can be either **quantitative** (i.e. numerical like weight or number of clients) or *qualitative* (i.e. an attribute like car colour or type of disease).

There are two types of quantitative variables:

(i) *Continuous.* A continuous variable is a variable which could take all possible values within a given range, e.g. the height of a tree.

(ii) *Discrete.* A discrete variable is a variable which increases in steps (often whole numbers), e.g. the number of rooms in a building.

A discrete variable does not have to consist only of whole numbers. For example, the size of shoes is also a discrete variable, and the sizes go up in steps of a half (5, $5\frac{1}{2}$, 6, $6\frac{1}{2}$, etc.).

The number of steps climbed is a *discrete* variable.

The distance travelled on the escalator is a *continuous* variable.

**Observation.** An observation is the value taken by a variable. For example, an age of 17 years is an observation when the variable is age.

**Population.** The term *'population'* means everything (or everybody) in the category you are considering. For example, if you were making a study of cathedrals, the population could be all the cathedrals in Britain. If you were investigating what attracts people to certain types of holiday, the population would be all holiday-makers.

# 16 Classification and Tabulation of Data

## 16.1 Tabulation

The purpose of tabulation is to arrange information, after collection and classification, into a compact space so that it can be read easily and quickly. It then may be represented pictorially to enable relevant facts to be seen readily, as explained in the next chapter.

Tabulation consists of entering the data found in columns or rows.

**EXAMPLE**

The numbers of pensioners living in certain villages were:

| Village | Number of pensioners |
|---|---|
| Ashurst | 31 |
| Botleigh | 17 |
| Crow | 28 |
| Downton | 24 |
| Eaglecliffe | 19 |
| Fillingdales | 33 |
| Total | 152 |

It is important that the tables produced are neat, all rows and columns are clearly identified, and that units (where appropriate) are given.

## 16.2 Classification of data

Assuming that additional data had been collected, more detailed information could be given by subdividing the rows and/or columns.

**EXAMPLE**

Using the data from the Example above and subdividing the columns into male and female gives more information about the pensioners:

| Village | Number of pensioners | |
|---|---|---|
| | Male | Female |
| Ashurst | 12 | 19 |
| Botleigh | 5 | 12 |
| Crow | 10 | 18 |
| Downton | 11 | 13 |
| Eaglecliffe | 9 | 10 |
| Fillingdales | 15 | 18 |
| Total | 62 | 90 |

*Note:* It is essential that all the relevant information is collected during the survey.

It is not possible, for example, to determine a person's sex after the survey has been completed.

## 16.3 Tally charts

It is common to record the data by means of a **tally chart**.

Suppose a survey was being carried out to determine the popularity of the various activities offered at a local leisure centre. First a list would be drawn up of possible activities: swimming, badminton, fitness training, etc. Then each person entering the leisure centre would be asked which activity they were paying for and a tally mark ( *I* ) would be recorded against the chosen activity.

To enable the results to be totalled quickly, it is usual to tally in groups of five, the fifth stroke being drawn diagonally across the previous four: ~~IIII~~.

A section of the results for this survey could look like this:

| Activity | Tally | Total |
|---|---|---|
| Archery | ~~IIII~~ ~~IIII~~ I | 11 |
| Badminton | ~~IIII~~ ~~IIII~~ ~~IIII~~ ~~IIII~~ III | 23 |
| Bowls | ~~IIII~~ III | 8 |
| Fitness room | ~~IIII~~ ~~IIII~~ ~~IIII~~ IIII | 19 |
| Judo | ~~IIII~~ ~~IIII~~ ~~IIII~~ ~~IIII~~ | 20 |

## 16.4 Frequency tables

A table which shows a set of variables and the number of times each variable occurs (its **frequency**) is called a **frequency table** or **frequency distribution table**.

If a large amount of quantitative data has been collected, it is generally convenient to record the information in a more compact form by combining variables into **groups** or **classes**. Continuous variables, such as time, length, speed, will normally be grouped before the information is collected.

Suppose the leisure centre survey is extended to find the amount of time people spend in the centre.

First the size of each class is decided (say 15 minutes).

Then a table is drawn up of all the classes.

The time each person in the survey has spent in the centre is tallied against the appropriate class and hence the frequencies are found.

| Time spent (minutes) | Tally | Frequency |
|---|---|---|
| Less than 15 | ~~IIII~~ I | 6 |
| 15–29 | ~~IIII~~ ~~IIII~~ | 10 |
| 30–44 | ~~IIII~~ ~~IIII~~ III | 13 |
| 45–59 | ~~IIII~~ ~~IIII~~ II | 12 |
| 60–74 | ~~IIII~~ ~~IIII~~ ~~IIII~~ I | 16 |
| 75–89 | ~~IIII~~ ~~IIII~~ ~~IIII~~ ~~IIII~~ | 20 |
| 90–104 | ~~IIII~~ ~~IIII~~ ~~IIII~~ ~~IIII~~ I | 21 |
| 105–119 | ~~IIII~~ ~~IIII~~ ~~IIII~~ II | 17 |
| 120–134 | ~~IIII~~ ~~IIII~~ IIII | 14 |

## EXERCISE 16.1

**1** During a survey to find how knowledgeable the general public is about art, 40 people were asked to name as many artists as possible in one minute. The responses were:

| | | | | | | | |
|---|---|---|---|---|---|---|---|
| 1 | 5 | 3 | 5 | 1 | 8 | 15 | 1 |
| 2 | 2 | 1 | 3 | 4 | 4 | 1 | 2 |
| 13 | 11 | 8 | 6 | 1 | 2 | 5 | 2 |
| 3 | 4 | 9 | 2 | 3 | 10 | 1 | 6 |
| 2 | 7 | 1 | 4 | 6 | 4 | 5 | 3 |

Use a tally chart to draw up a frequency table for this data.

**2** A textile mill spins yarn. The thickness of the yarn is measured at intervals, and the measurements, in millimetres, of a sample of 50 are given below.

| | | | | | | |
|---|---|---|---|---|---|---|
| 0.72 | 0.98 | 0.81 | 0.96 | 0.91 | 0.90 | 0.76 |
| 0.92 | 0.95 | 0.91 | 0.83 | 0.91 | 0.89 | 0.86 |
| 0.93 | 0.94 | 0.78 | 0.93 | 0.83 | 0.86 | 0.91 |
| 0.78 | 0.92 | 0.88 | 1.03 | 1.04 | 1.01 | 0.94 |
| 1.03 | 0.90 | 0.85 | 0.85 | 0.91 | 0.82 | 0.88 |
| 0.95 | 1.02 | 0.99 | 0.97 | 0.92 | 0.82 | 0.90 |
| 1.03 | 0.93 | 0.94 | 0.96 | 0.87 | 0.93 | 0.89 |
| 0.92 | | | | | | |

Using intervals of 0.70–0.74, 0.75–0.79, 0.80–0.84, etc., draw up a tally chart to obtain the frequency distribution.

**3** A small business carried out a survey to find the number of days absence of the employees over one year. The results were:

| | | | | | | | | |
|---|---|---|---|---|---|---|---|---|
| 3 | 1 | 4 | 2 | 1 | 15 | 20 | 5 | 15 |
| 0 | 17 | 26 | 0 | 11 | 1 | 3 | 10 | 8 |
| 15 | 10 | 17 | 10 | 6 | 13 | 12 | 10 | 14 |
| 5 | 8 | 3 | 21 | 0 | 3 | 18 | 3 | 18 |
| 3 | 42 | 9 | 18 | 10 | 21 | 10 | 5 | 6 |
| 14 | 1 | 5 | 5 | 0 | 5 | 7 | 30 | 9 |
| 0 | 5 | 6 | 25 | 23 | 6 | 4 | 11 | 12 |

Collect the information on a frequency table using intervals 0–4, 5–9, 10–14, 15–19, etc.

**4** During a survey into changes in the conditions of work of clerical staff, 50 workers gave their present salaries (in £) as:

| | | | | |
|---|---|---|---|---|
| 14030 | 12670 | 10180 | 11320 | 9870 |
| 10120 | 10130 | 15460 | 13680 | 9920 |
| 13830 | 11610 | 11880 | 14280 | 12200 |
| 11020 | 11570 | 10990 | 9700 | 11810 |
| 10880 | 11370 | 12090 | 9800 | 9670 |
| 12230 | 11680 | 8590 | 9680 | 10280 |
| 10420 | 12120 | 9330 | 10540 | 7490 |
| 9240 | 8990 | 7630 | 11010 | 9180 |
| 8320 | 8640 | 15200 | 8680 | 12040 |
| 8680 | 7480 | 7720 | 8290 | 8470 |

Organize the data on a frequency table using intervals £7000–£7999, £8000–£8999, etc.

**5** The staff in a medical practice monitored the waiting times of patients from the time the patient sat down until called to see the doctor. The times, in minutes and seconds, were:

| | | | | |
|---|---|---|---|---|
| 10:03 | 12:05 | 7:15 | 9:44 | 11:15 |
| 10:02 | 14:23 | 12:15 | 12:42 | 15:00 |
| 5:43 | 9:08 | 9:53 | 9:03 | 14:21 |
| 7:24 | 10:57 | 12:26 | 7:13 | 15:30 |
| 10:53 | 12:26 | 10:57 | 13:48 | 8:00 |
| 9:24 | 12:48 | 10:17 | 11:02 | 7:48 |
| 9:56 | 14:09 | 7:23 | 9:03 | 9:59 |
| 9:32 | 8:05 | 7:53 | 14:23 | 13:03 |

**a** Write each waiting time to the nearest minute and use a tally chart to obtain the frequencies.

**b** Summarise the given waiting times in a frequency table using intervals $5.00 \leqslant T < 7.00$, $7.00 \leqslant T < 9.00$, etc., where $T$ is the waiting time.

**6** The weights (in kg) of 63 male patients admitted to a ward were recorded as:

| | | | | | | |
|---|---|---|---|---|---|---|
| 72.4 | 68.2 | 69.3 | 71.1 | 66.8 | 67.2 | 65.4 |
| 68.0 | 78.9 | 76.0 | 70.8 | 64.3 | 82.3 | 70.2 |
| 74.2 | 76.7 | 65.5 | 71.6 | 74.1 | 68.7 | 66.8 |
| 83.3 | 74.9 | 71.5 | 75.7 | 71.6 | 73.2 | 82.5 |
| 73.8 | 78.2 | 65.6 | 76.9 | 76.8 | 81.5 | 77.2 |
| 75.8 | 75.4 | 80.3 | 78.0 | 68.3 | 76.0 | 78.5 |
| 78.8 | 71.7 | 74.4 | 69.8 | 77.6 | 73.4 | 77.3 |
| 74.9 | 72.4 | 66.9 | 73.7 | 74.4 | 68.8 | 82.6 |
| 73.7 | 79.8 | 74.0 | 71.8 | 73.4 | 76.0 | 79.2 |

Using intervals 60–, 65–, 70–, etc., summarise the information in a frequency table.

# 17 Statistics on Display

## 17.1 Pictorial representation of data

The presentation of data in the form of tables has been considered in Unit 16. However, most people find that the presentation of data is more effective, and easier to understand, if the data is presented in a pictorial or diagrammatic form.

The pictorial presentation used must enable the data to be more effectively displayed and more easily understood. The diagrams must be fully labelled, clear and should not be capable of visual misrepresentation. Types of pictorial representation in common use are the pictogram, bar chart, pie chart and frequency polygon.

Statistical packages may also be used to present data in a variety of ways. You cannot, however, rely completely on a computer to produce your pie charts, pictographs, etc. You must also be able to carry out the necessary calculations yourself and draw the most appropriate diagrams for the given data.

There are many ways of presenting data in pictorial form. It is clearly necessary to be able to interpret correctly any diagrams given.

The general interpretation of statistical pictures and graphs is that the bigger the representation, the larger the population in that group. However, it is also possible to interpret statistical diagrams so as to be able to calculate the population of each group.

## 17.2 Pictograms

In a **pictogram** data is represented by the repeated use of a pictorial symbol. The example below shows how a pictogram works.

**EXAMPLE**

A survey of 1000 people living in Freeton was taken, to see what colour of cars they owned. Represent this data in the form of a pictogram. The results of the survey were:
Colour of cars Number of cars

| Colour of cars | Number of cars |
|---|---|
| Red | 60 |
| White | 200 |
| Blue | 100 |
| Grey | 80 |
| Gold | 50 |
| Black | 30 |

Key: [car symbol] = 20 cars

Here is one possibility. A full car symbol represents 20 cars; half a car represents 10 cars. It is not possible to show small fractions of a symbol accurately, and the detail required should not normally be to more than half of a symbol (but certain symbols may allow for a quarter).

## EXERCISE 17.1

**1** The numbers of bottles of champagne sold in five villages is shown on the following pictogram:

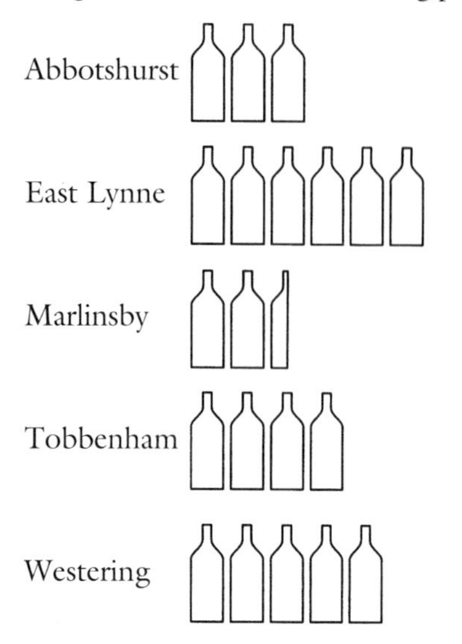


**a** In which village were the most bottles sold?
**b** How many bottles were sold in Martinsby?
**c** How many bottles were sold in total?

**2** The contents of a fruit bowl comprised:

| | | | |
|---|---|---|---|
| Apples | 7 | Bananas | 3 |
| Pears | 5 | Peaches | 7 |
| Kiwi fruit | 6 | Oranges | 2 |

Illustrate this data by means of a pictogram.

**3** Students in a department of a college were asked about the type of accommodation in which they lived. The data was:

| | | | |
|---|---|---|---|
| Flat | 25 | Semi-detached house | 40 |
| Maisonette | 5 | Detached house | 30 |

Illustrate this data by means of a pictogram.

**4** In a survey to find the most popular design on Christmas cards, 600 people were asked which animal they preferred. The results are shown on the pictogram.

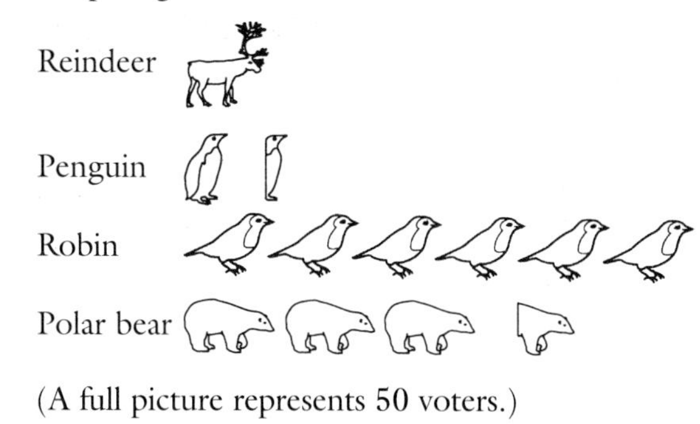


(A full picture represents 50 voters.)

**a** How many people chose the polar bears design?
**b** What percentage of people chose the reindeer?
**c** Find the ratio of the votes for robins to the votes for penguins.

**5** The number of people present in a Paris fashion show were:

| | |
|---|---|
| Individual buyers | 54 |
| Store buyers | 18 |
| Celebrities | 27 |
| Photographers | 45 |
| Journalists | 36 |

Illustrate this data by means of a pictogram.

**6** The numbers of employees in four solicitors' offices were:

| | |
|---|---|
| Archibald and Archibald | 8 |
| Dugdale, Wynne and Luff | 10 |
| JSC Weston-Hough | 6 |
| Mordecai and Sons | 12 |

Draw a pictogram to represent this data.

**7** A college canteen carried out a survey to decide which type of bread roll to serve. The answers are shown in the pictogram.

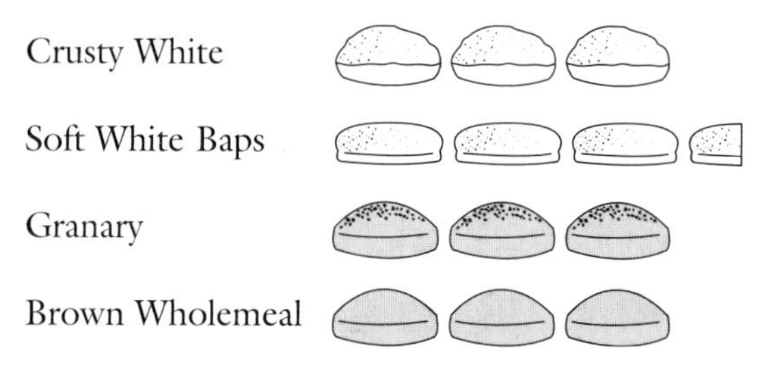


(One roll represents 4 votes.)

**a** Which was the favourite type of roll?
**b** How many customers preferred wholemeal rolls?
**c** What percentage preferred white bread?

**8** The number of patients on a register of six doctors in a group practice is:

| | |
|---|---|
| Dr Smith | 2400 |
| Dr Rawlings | 1800 |
| Dr Wong | 3000 |
| Dr Payne | 2700 |
| Dr Williams | 23100 |
| Dr Fisher | 1500 |

Show this information on a pictogram.

## 17.3 Bar charts

A **bar chart** is a diagram consisting of columns (i.e. bars), the heights of which indicate the frequencies. Bar charts may be used to display discrete or qualitative data.

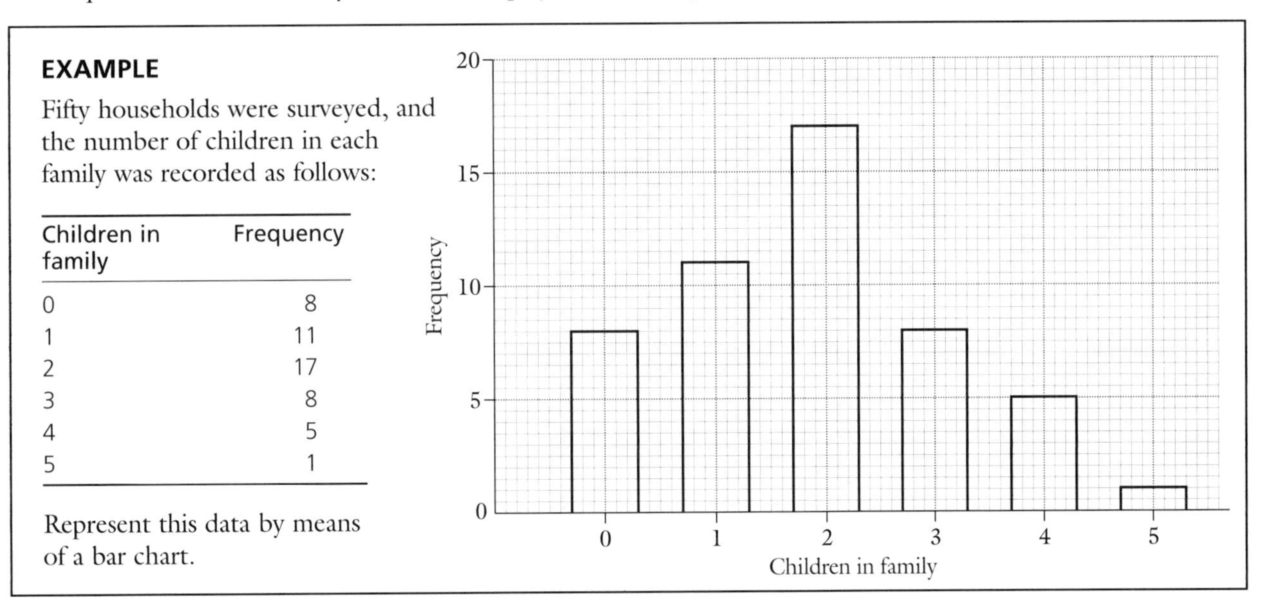


**EXAMPLE**

Fifty households were surveyed, and the number of children in each family was recorded as follows:

| Children in family | Frequency |
|---|---|
| 0 | 8 |
| 1 | 11 |
| 2 | 17 |
| 3 | 8 |
| 4 | 5 |
| 5 | 1 |

Represent this data by means of a bar chart.

### Dual bar charts

Dual bar charts are used when two different sets of information are given on connected topics.

**EXAMPLE**

The number of people over 17 years old, and the number of people holding driving licences in a particular street were found over a period of years.
These are as shown below.
Represent this data by means of a dual bar chart.

| Year | 1995 | 1996 | 1997 | 1998 | 1999 | 2000 |
|---|---|---|---|---|---|---|
| No. of people over 17 | 32 | 27 | 29 | 31 | 33 | 39 |
| No. of people with driving licence | 12 | 17 | 19 | 11 | 24 | 28 |

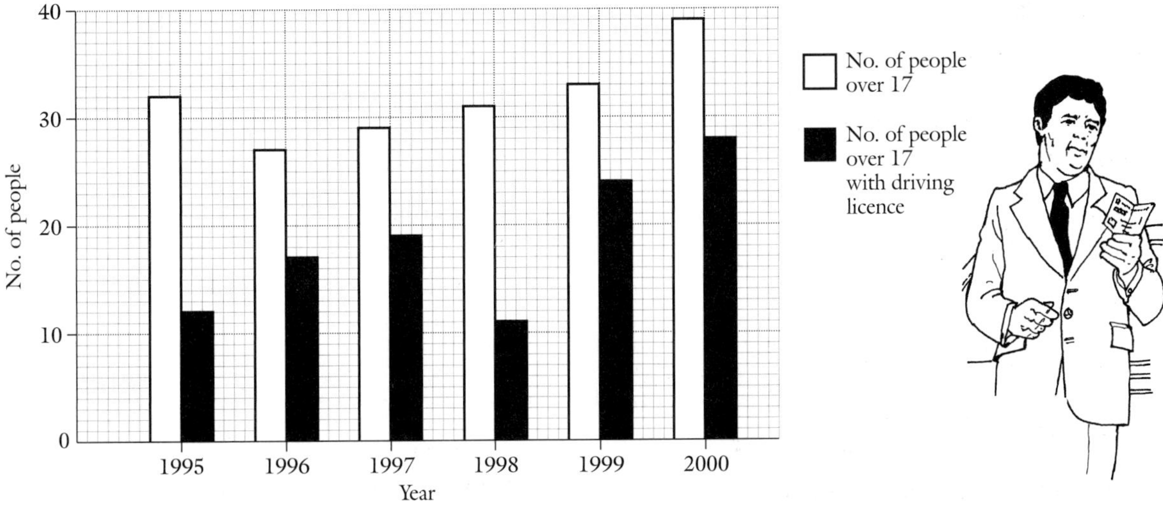

## Sectional bar charts

Sectional bar charts, or **component bar charts**, are used when two, or more, different sets of information are given on the same topics. They are particularly useful when the *total* of the two or more bars is also of interest.

### EXAMPLE

The numbers of saloons and hatchbacks sold by a garage were recorded.

| Month | Jan | Feb | Mar | Apr | May | Jun |
|---|---|---|---|---|---|---|
| Saloons | 18 | 7 | 8 | 12 | 10 | 13 |
| Hatchbacks | 16 | 12 | 9 | 7 | 9 | 8 |

Represent this data by means of a sectional bar chart.

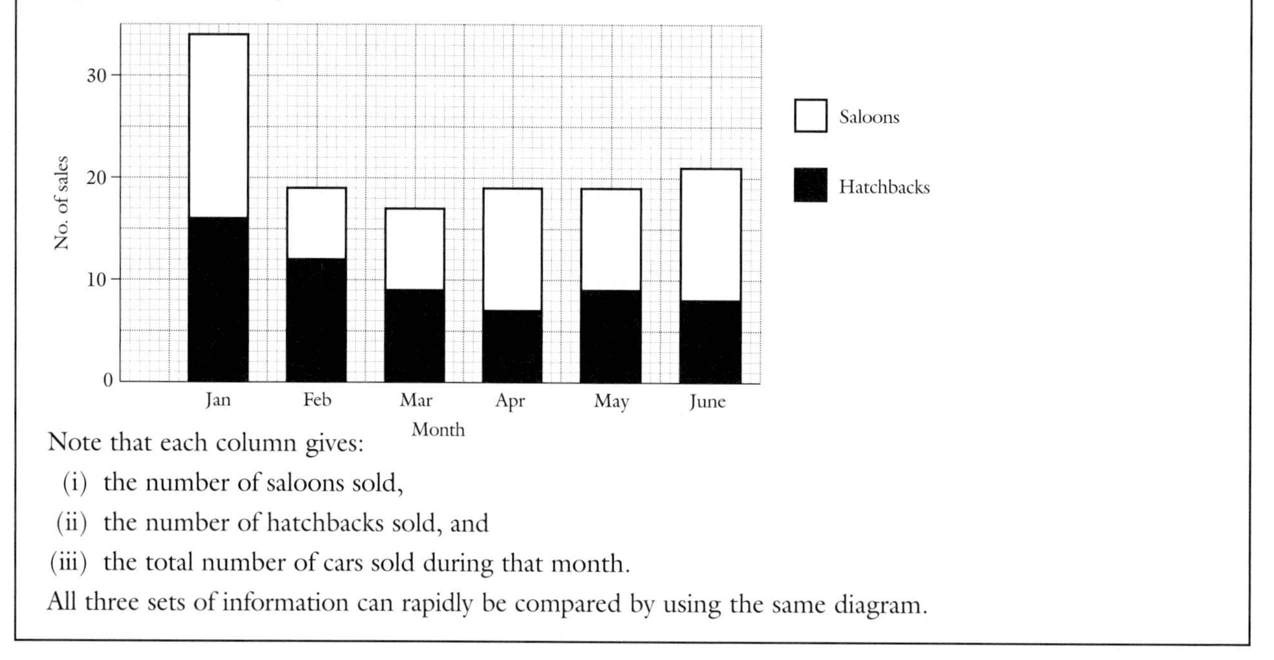


Note that each column gives:

(i) the number of saloons sold,

(ii) the number of hatchbacks sold, and

(iii) the total number of cars sold during that month.

All three sets of information can rapidly be compared by using the same diagram.

### EXERCISE 17.2

1 The bar chart shows the type of trees recorded during a survey of a section of a forest.

Oak
Elm
Chestnut
Beech
Conifer
Cedar

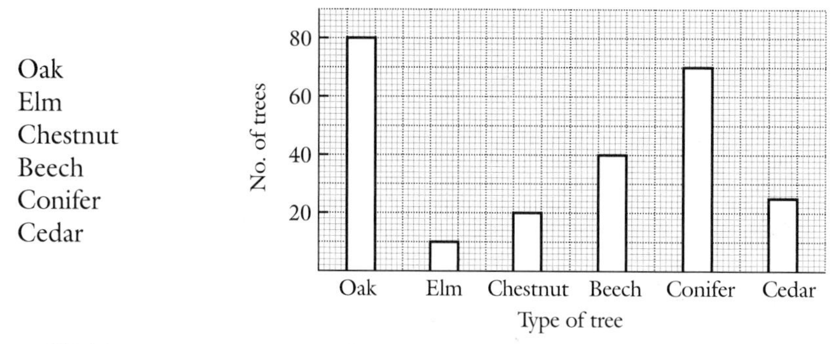


a Which tree was seen most frequently?

b Reconstruct the frequency table.

c What was the total number of trees growing in this area?

2 Twenty people noted the television channel they were watching at 8.15pm on two successive nights. The results were:

| | First night | Second night |
|---|---|---|
| BBC 1 | 6 | 7 |
| BBC 2 | 4 | 1 |
| ITV | 6 | 6 |
| CHANNEL 4 | 3 | 4 |
| SATELLITE | 1 | 2 |

Draw a suitable bar chart to illustrate this data.

3 Thirty students designed ball gowns. The colours of the gowns were 12 black, 5 gold, 10 red, 2 blue, and 1 green.

Illustrate this information by means of a bar chart.

4 A graphics company investigated how many hours their employees actually worked on a computer during a week in January 1998 and again in 1999. The results were:

| Time per week (hours): | 0 | 0–1 | 1–2 | 2–4 | 4–10 | 10–20 | 20–40 |
|---|---|---|---|---|---|---|---|
| 1998 | 32 | 17 | 4 | 10 | 2 | 1 | 25 |
| 1999 | 17 | 2 | 5 | 11 | 12 | 21 | 34 |

Represent this data by means of a dual bar chart.

5 The number of students in an art class on six successive evenings were:

| | Eve 1 | Eve 2 | Eve 3 | Eve 4 | Eve 5 | Eve 6 |
|---|---|---|---|---|---|---|
| Male | 11 | 9 | 8 | 9 | 7 | 6 |
| Female | 7 | 8 | 9 | 10 | 11 | 13 |

Illustrate this information by means of a section bar chart.

6 In a three-month period, the number of days in which different products were advertised on two hoardings were compared. These are shown in the dual bar charts below.

a One hoarding was in an inner city, and the other one was in a suburban area.
Which hoarding was in the inner city?

b How many more days did Hoarding A advertise alcohol than Hoarding B?

c Which products were advertised only on Hoarding A?

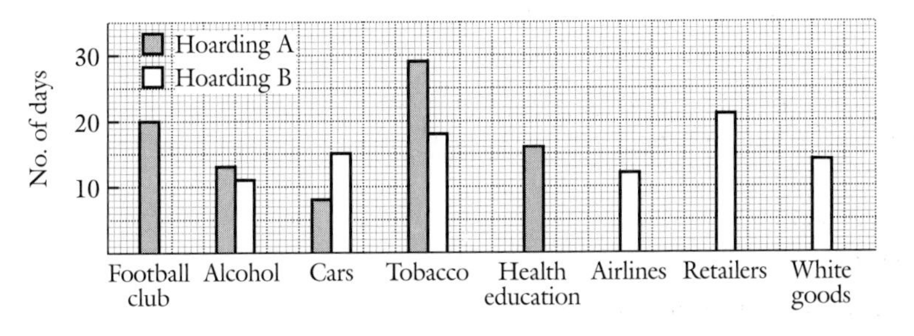

7 The numbers of large appliances sold by an electrical retailer during one day were:

| | |
|---|---|
| Washing machines | 35 |
| Cookers | 28 |
| Televisions | 38 |
| Video Recorders | 12 |
| Refrigerators | 21 |

Illustrate this data by means of a bar chart.

8 An insurance company keeps records of life endowment policies sold by its representatives. In a six-month period the number of policies sold by their top salesmen were:

| | With Profits | Without Profits |
|---|---|---|
| January | 11 | 21 |
| February | 15 | 12 |
| March | 28 | 13 |
| April | 21 | 20 |
| May | 18 | 27 |
| June | 16 | 30 |

Represent this data by means of a section bar chart.

9 The number of houses sold by four agents in the first six months of 1998 and 1999 were:

| | 1998 | 1999 |
|---|---|---|
| John | 27 | 29 |
| Mary | 15 | 24 |
| Carl | 28 | 19 |
| Latha | 22 | 32 |

Represent this data by means of a dual bar chart.

10 The goals scored in 42 football league matches on Saturday 26 March 2000 were:

| Number of goals in match | 0 | 1 | 2 | 3 | 4 | 5 | 6 |
|---|---|---|---|---|---|---|---|
| Number of matches | 1 | 9 | 10 | 12 | 6 | 3 | 1 |

Illustrate this information by means of a bar chart.

11 Jean-Paul and Michelle noted how the cost of their summer holiday had varied in the last two years. The money had been spent as shown:

| | Travel | Rent of Villa | Food | Drink | Entertainment | Insurance |
|---|---|---|---|---|---|---|
| Cost 1998 (£) | 720 | 820 | 550 | 140 | 250 | 110 |
| Cost 1999 (£) | 650 | 920 | 480 | 160 | 150 | 120 |

Represent this data by means of a dual bar chart.

**12** The percentage of trade of certain countries with the UK and the USA was calculated in 1996 to be as follows:

| | Barbados | Canada | Hong Kong | Ireland | Luxembourg | Sweden |
|---|---|---|---|---|---|---|
| % trade with E.C. | 16.8 | 1.4 | 5.0 | 25.7 | 3.1 | 9.6 |
| % trade with USA | 13.5 | 82.3 | 25.4 | 8.2 | 6.3 | 8.3 |

Illustrate this information by means of a dual bar chart.

**13** The number of employees in five small printing companies were:

| | | |
|---|---|---|
| Firm A | 6 secretaries | and 17 other staff |
| Firm B | 8 secretaries | and 14 other staff |
| Firm C | 11 secretaries | and 10 other staff |
| Firm D | 5 secretaries | and 12 other staff |
| Firm E | 7 secretaries | and 6 other staff |

Represent this data by means of a sectional bar chart.

## 17.4 Pie charts

A **pie chart** is another type of diagram for displaying information. It is particularly suitable if you want to illustrate how a population is divided up into different parts and what proportion of the whole each part represents. The bigger the proportion, the bigger the slice (or 'sector').

**EXAMPLE**

Represent by a pie chart the following data.

The mode of transport of 90 students into college was found to be:

| | |
|---|---|
| Walking | 12 |
| Cycling | 8 |
| Bus | 26 |
| Train | 33 |
| Car | 11 |
| Total | 90 |

Represent this data by means of a pie chart.

A circle has 360°. Divide this by 90 to give 4°. This is then the angle of the pie chart that represents each individual person.

Since 12 people walk to college, they will be represented by $12 \times 4° = 48°$.

Similarly for the others:

| | Angle in pie chart |
|---|---|
| Walking | $12 \times 4 = 48°$ |
| Cycling | $8 \times 4 = 32°$ |
| Bus | $26 \times 4 = 104°$ |
| Train | $33 \times 4 = 132°$ |
| Car | $11 \times 4 = 44°$ |
| | Total = 360° |

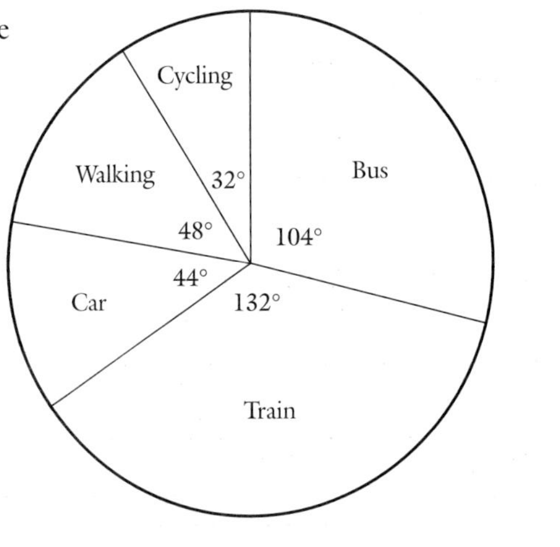

*Method for calculating the angles on a pie chart*

Here is a summary of how to work out the size of each bite of 'pie'.

(i) Add up the frequencies. This will give you the total population (call it $p$) to be represented by the pie.

(ii) Divide this number into 360.

(iii) Multiply each individual frequency by this result. This will give you the angle for each section of the pie chart.

## Interpreting pie charts

The initial interpretation is the fact that the largest portion of a pie chart relates to the largest group, and the smallest portion to the smallest group. However, if any of the data is known, the rest of the data can be calculated.

**EXAMPLE 1**

The pie chart below shows the number of students in different sections of a college. 220 students are in the Construction department.

a How many students are there in the college?

b How many students are there in Catering?

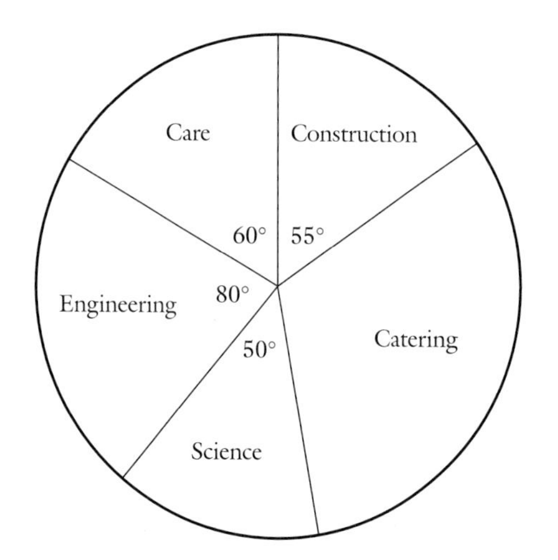


a 55° represents 220 students.

$\therefore$ 1° represents $\frac{220}{55} = 4$ students.

The complete circle (360°) represents $4 \times 360 = 1440$ students.

$\therefore$ There are 1440 students in the college.

b The angle representing Catering is $360 - (80 + 55 + 60 + 50) = 115°$.

$\therefore$ The number of students in Catering is $4 \times 115 = 460$.

## EXERCISE 17.3

Illustrate the data given in questions 1, 2 and 3 below by means of a pie chart.

1 The types of central heating used by households in a village were:

| | |
|---|---|
| Solid fuel | 14 |
| Gas | 105 |
| Electricity | 41 |
| None | 20 |

2 240 students were asked what they were intending to do during next year. The results were:

80 going to university
86 staying at college
64 going into employment
10 with no firm intention.

3 The numbers of bedrooms in 720 houses recorded as:

| | |
|---|---|
| 1 bedroom | 80 |
| 2 bedrooms | 235 |
| 3 bedrooms | 364 |
| 4 bedrooms | 39 |
| 5 bedrooms | 2 |

4 The pie chart shows the different drinks sold at lunchtime in a college. 720 drinks were sold in total.

Find the number of each different drink sold during lunchtime:

**a** Coke
**b** Orange
**c** Coffee
**d** Chocolate

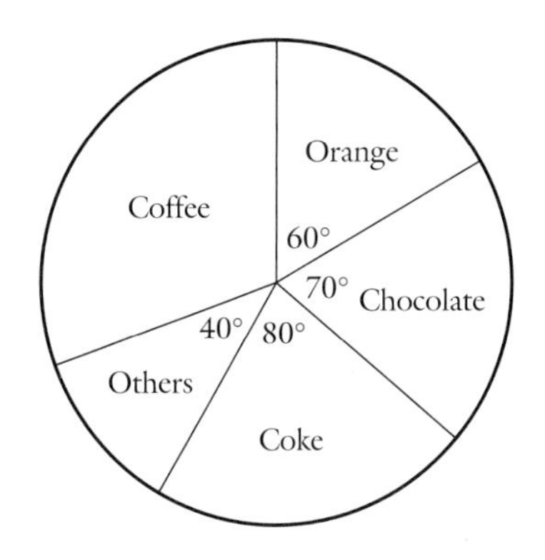


5 The number of special birthday cards sold by a newsagents in a week were:

| | |
|---|---|
| Mum | 42 |
| Dad | 34 |
| Grandad | 18 |
| Granny | 16 |
| Brother | 25 |
| Sister | 17 |
| Son | 15 |
| Daughter | 13 |

Illustrate this information on a pie chart.

6 An artist designed book jackets for 225 books during a five-year period with one publishing house. The books were classified as Thriller, Romance, Travel, Hobby and Science.

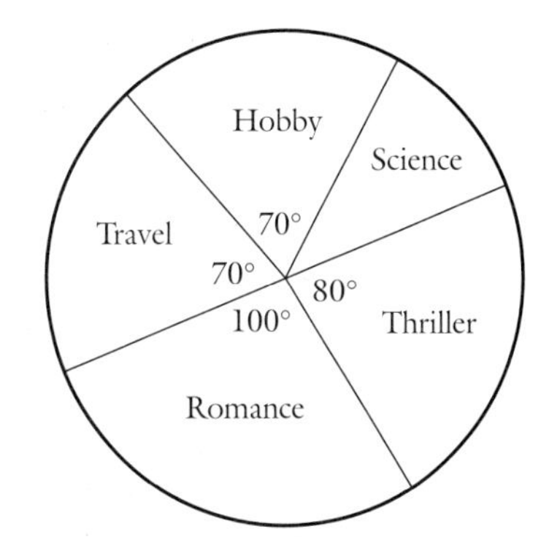


**a** What size angle represents science books?
**b** How many jacket designs did the artist create for science books?
**c** How many jacket designs were created for thrillers?
**d** What fraction of the designs were for romances? (Give your answer in its lowest terms.)

7 Each pound spent at the Winchester Theatre Royal box-office is used to meet the theatre's expenses as follows:

| | |
|---|---|
| Performance fees | 60p |
| Salaries | 17p |
| Premises and depreciation | 10p |
| Administration | 5p |
| Publicity | 5p |
| Equipment | 3p |

Draw a pie chart to show how each pound is spent.

**8** At a sports centre, the ages of 100 people were recorded as follows:

| | |
|---|---|
| Under 20 years | 30 |
| 20 to 29 years | 15 |
| 30 to 39 years | 12 |
| 40 to 59 years | 14 |
| 60 years and over | 29 |

Construct a pie chart to illustrate this data.

**9** The pie chart shows the number of passengers flying from London to Miami on one afternoon. 1800 passengers in total flew this route on that afternoon.

Find the number flying:

**a** Virgin

**b** American Airlines

**c** British Airways

The plane used by Virgin is a Boeing 747 seating 370 passengers.

**d** What percentage of the Virgin seats were occupied?

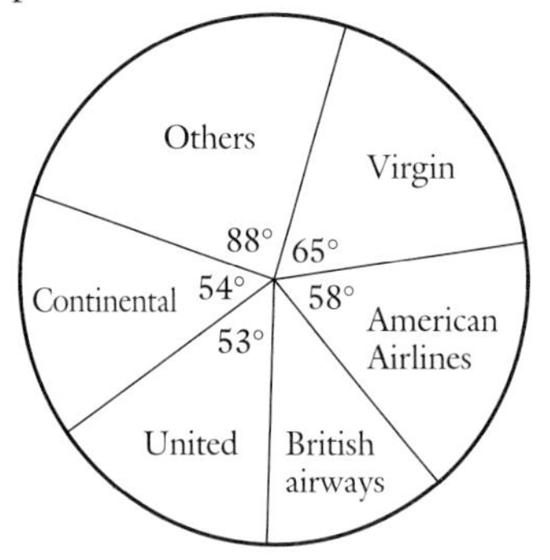


**10** Out of 90 employees in a manufacturing company, there were 10 managers, 15 salesmen, 52 production line workers, 4 typists, and 9 quality controllers.
Illustrate this information on a pie chart.

**11** A manufacturer of combined harvesters commissioned a survey on the use of agricultural land in South Australia. The result is shown in the pie chart.

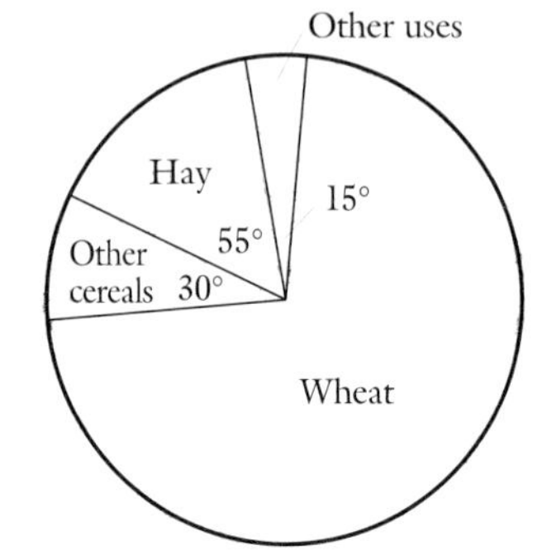


**a** Find the percentage of land use for wheat.

**b** The total acreage is 3 000 000 acres. What area was used for hay?

**14** The holiday destinations of 60 people were:

| France | Spain | Greece | Tunisia | USA | Caribbean | Portugal |
|---|---|---|---|---|---|---|
| 21 | 15 | 6 | 3 | 8 | 5 | 2 |

Represent this information by means of a pie chart.

## 17.5 Line graphs

A bar chart can be replaced by a line graph, provided that the quantity on the horizontal axis is continuous, e.g. age, temperature or time.

In this case the data is plotted as a series of points which are joined by straight lines.

Line graphs associated with time are called **time-series graphs**.

They are used, for example, by geographers to illustrate monthly rainfall or yearly crop yield, etc., and by businesses to display information about profits or production over a period of time.

They show trends and have the advantage that they can be easily extended.

**EXAMPLE**

The temperatures, recorded every six hours, of a patient in a hospital ward are given on the table:

| | Mon. | | | | Tues. | | | | | Wed. |
|---|---|---|---|---|---|---|---|---|---|---|
| Time (hours) | 06 | 12 | 18 | 00 | 06 | 12 | 18 | 00 | 06 | 12 |
| Temperature (°F) | 99.0 | 99.12 | 99.12 | 99.2 | 99.2 | 98.99 | 98.68 | 98.6 | 98.6 | |

Represent this data by means of a line graph.

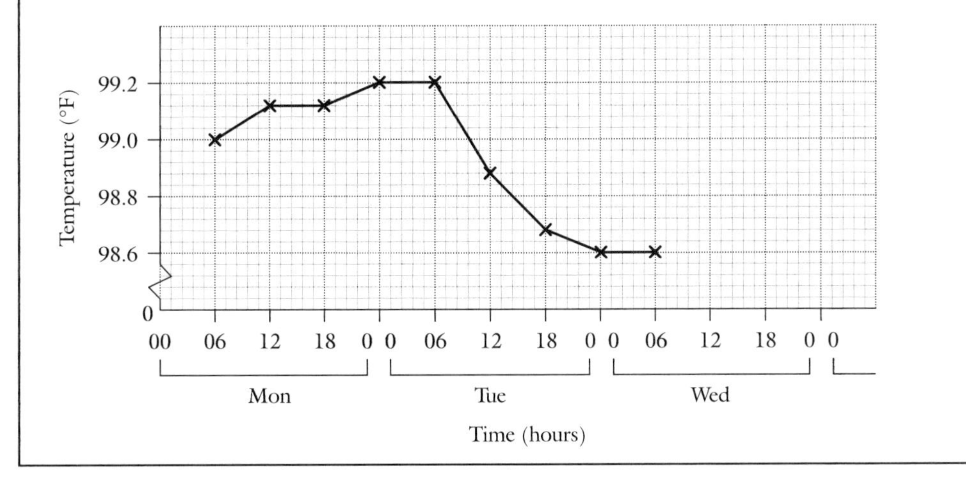


## EXERCISE 17.4

Illustrate the data given in questions 1, 2 and 3, using a line graph.

**1** The rainfall during a period of six months was:

| Month | Jan | Feb | Mar | Apr | May | Jun |
|---|---|---|---|---|---|---|
| Rainfall (mm) | 75 | 192 | 86 | 89 | 25 | 19 |

**2** The maximum temperatures for six successive months at Sunbourne were:

| Month | Apr | May | Jun | Jul | Aug | Sep |
|---|---|---|---|---|---|---|
| Temperature (°C) | 61 | 74 | 72 | 91 | 85 | 56 |

**3** The girth of a tree was:

| Age (years) | 10 | 20 | 30 | 40 | 50 | 60 |
|---|---|---|---|---|---|---|
| Girth (cm) | 25 | 63 | 98 | 135 | 170 | 210 |

**4** In an art class, students were asked to draw a bowl of fruit. The number of grapes drawn by the students were:

| Number of grapes | 0 | 1 | 2 | 3 | 4 | 5 |
|---|---|---|---|---|---|---|
| Number of students | 2 | 7 | 6 | 3 | 2 | 1 |

Construct a line to show this information.

5 The graph below shows the number of births (measured along a vertical axis) in a given year (measured along a horizontal axis).
Answer the questions below by reading off the values from the graph.

a Estimate the number of children born in 1915.

b In which years were approximately 825 000 children born?

c Which year had the lowest number of births?

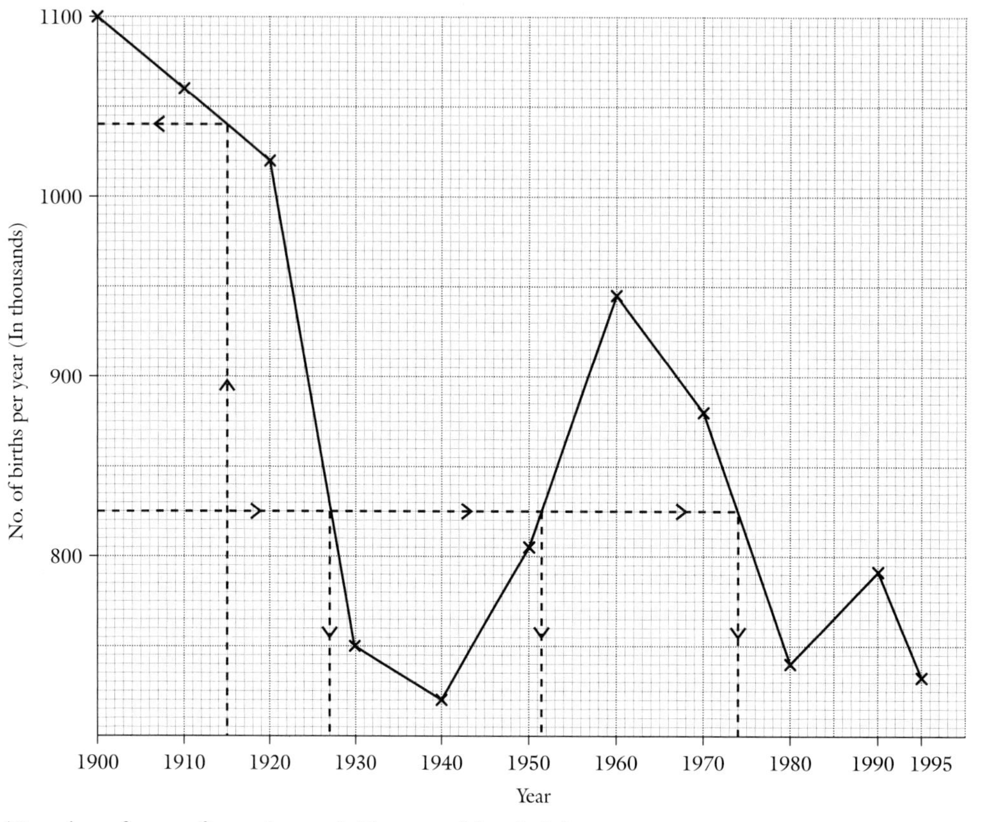


(Based on figures from *Annual Abstract of Statistics*)

6 Mordecai makes cuddly toys. The number of koala he made in one week is shown on the line graph below:

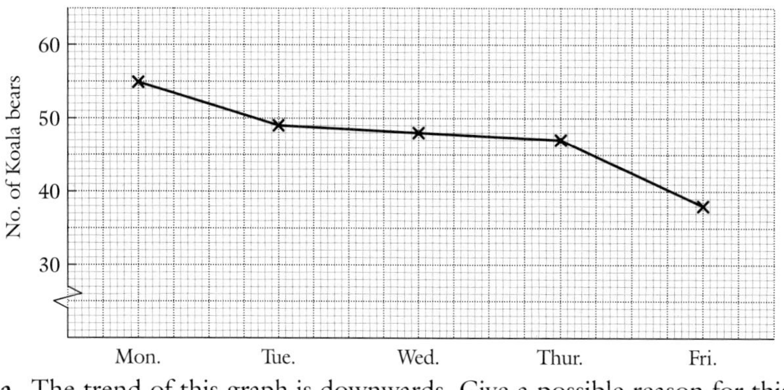


a The trend of this graph is downwards. Give a possible reason for this.

b During the following week, Mordecai's production of koala bears was:

| Mon | Tue | Wed | Thur | Fri |
|---|---|---|---|---|
| 41 | 46 | 49 | 48 | 42 |

What was the total production for each week?

7 The graph shows the median gross weekly earnings of women from 1976 to 1994.

a Estimate the median weekly earnings in 1981.

b Estimate the median weekly earnings in 1993.

c Why is the answer to **b** a better estimate than the answer to **a**?

d In which year did the average weekly wage reach £200?

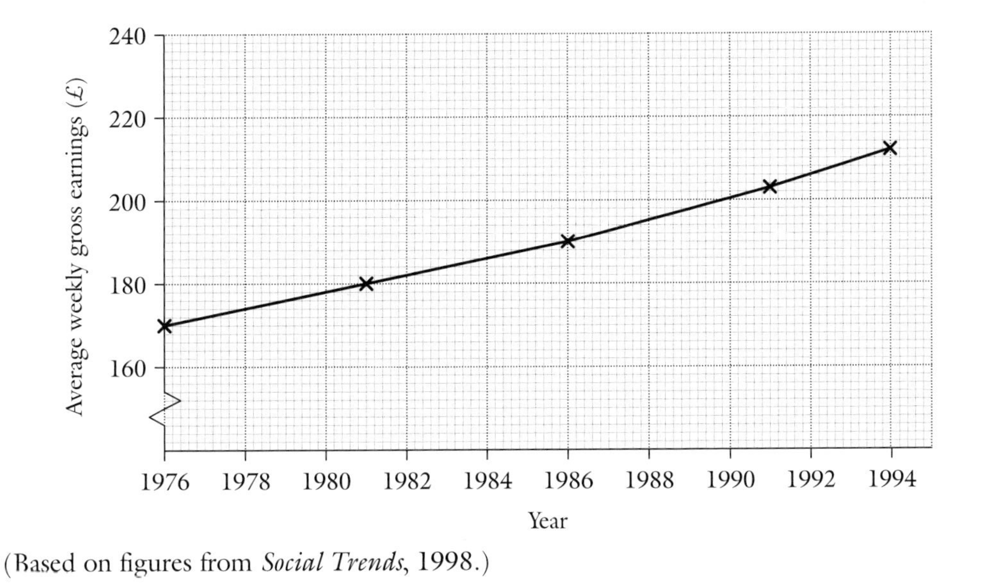


(Based on figures from *Social Trends*, 1998.)

## 17.6 Frequency polygons

A method of presenting data which is an alternative to a histogram is the **frequency polygon**. They are often used to compare frequency distributions, i.e. to compare the 'shapes' of the histograms, because it is possible to draw more than one frequency polygon on the same graph. It is easier to make comparisons using frequency polygons than using histograms.

For ungrouped data, the frequencies are plotted as points. For grouped data, which is more usual, the frequencies are plotted against the mid-point of the class interval. In both cases the points are joined with straight lines.

### EXAMPLE 1

The heights of 80 students were recorded. The data was:

| Height (cm) | 150–160 | 160–170 | 170–180 | 180–190 | 190–200 | 200–210 |
|---|---|---|---|---|---|---|
| No. of students | 4 | 7 | 15 | 47 | 6 | 1 |

Represent this data by means of a frequency polygon.

The new table is:

| Mid-point of class | 155 | 165 | 175 | 185 | 195 | 205 |
|---|---|---|---|---|---|---|
| Frequency density | 4 | 7 | 15 | 47 | 6 | 1 |

Frequency polygon

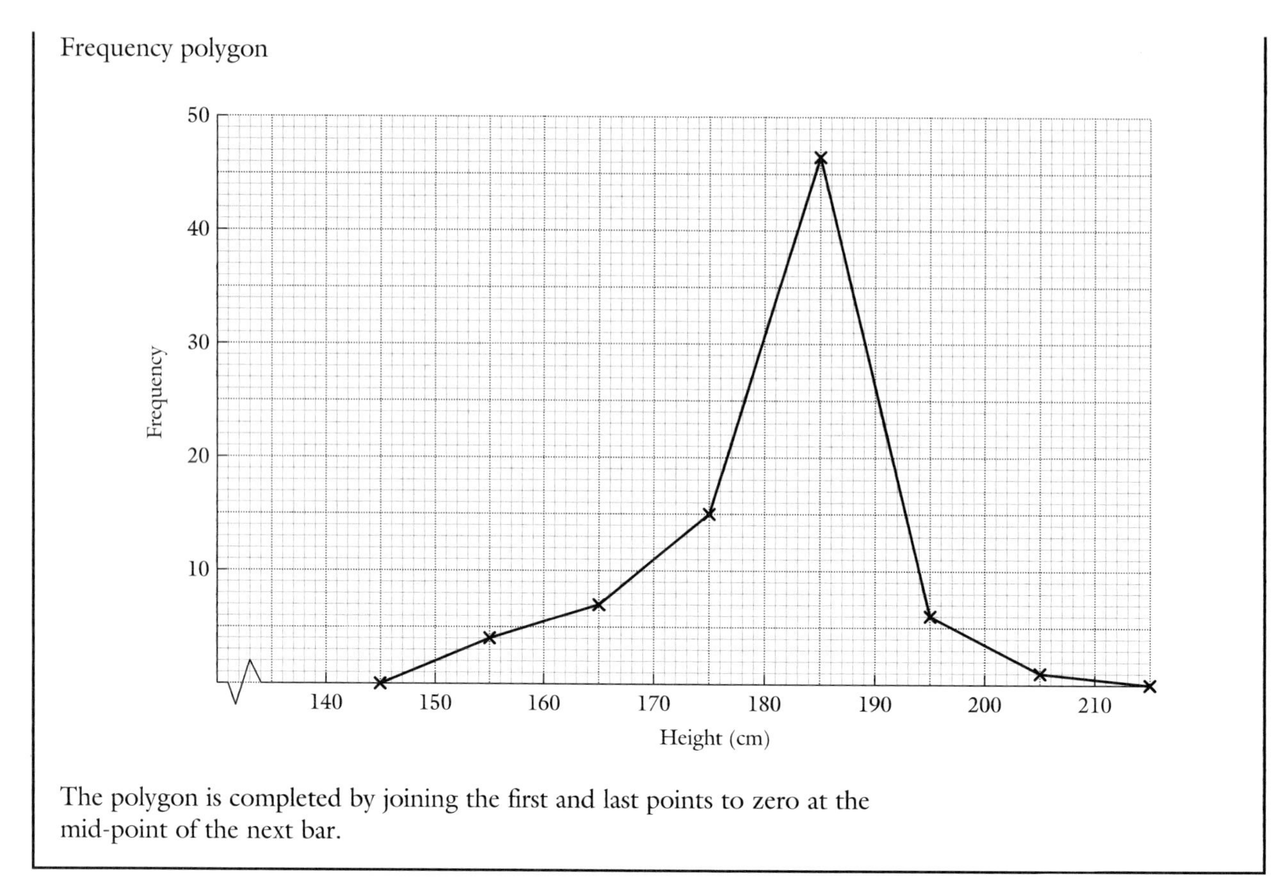


The polygon is completed by joining the first and last points to zero at the mid-point of the next bar.

When frequency polygons are used to compare two sets of data, it is the shapes of the distribution which are important. For this reason, the mid-points of the classes are often plotted against the actual frequencies, rather than the frequency densities which would be used for a histogram.

**EXAMPLE 2**

The mock examination results in Mathematics for two successive GCSE groups are recorded on the table below.

| Mark | 1–20 | 21–40 | 41–60 | 61–80 | 81–100 |
|---|---|---|---|---|---|
| Group 1 % frequency | 5 | 12 | 35 | 28 | 20 |
| Group 2 % frequency | 7 | 26 | 48 | 9 | 10 |

**a** Draw the frequency polygon for each group.

**b** Assuming the ability of the pupils was the same in each year, comment on the mock examination papers.

**a** In this example, the percentage frequencies are plotted against the class mid-points, which are 10.5, 30.5, 50.5, 70.5 and 90.5.

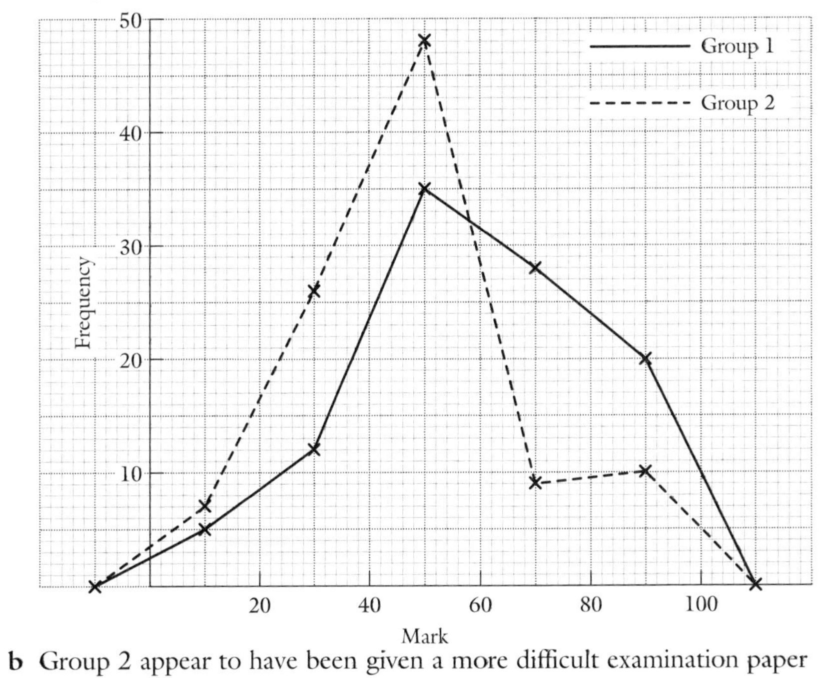


**b** Group 2 appear to have been given a more difficult examination paper than Group 1.

## EXERCISE 17.5

1 Draw the frequency polygon for the data given below.

The speeds of 100 cars on a motorway were recorded.
The data found was:

| Speed (mph) | 30–40 | 40–50 | 50–60 | 60–70 | 70–80 | 80–90 |
|---|---|---|---|---|---|---|
| Number of cars | 2 | 11 | 35 | 42 | 9 | 1 |

2 A teacher noted the rates of absence from her Maths class on Mondays and Fridays. The results are given on the table below.

| Number absent from class | 0 | 1 | 2 | 3 | 4 | 5 | 6 | 7 | 8 | 9 | 10 | 11 |
|---|---|---|---|---|---|---|---|---|---|---|---|---|
| Monday frequency | 3 | 6 | 6 | 7 | 4 | 4 | 3 | 0 | 0 | 0 | 0 | 0 |
| Friday frequency | 0 | 2 | 2 | 3 | 4 | 5 | 2 | 0 | 6 | 3 | 2 | 1 |

**a** Draw the frequency polygon for each day, using the same axes.
**b** Comment on the absence rates for the two days.

3 The table below shows the number of letters per word in 100 words of two books.

| No. of letters | 1 | 2 | 3 | 4 | 5 | 6 | 7 | 8 | 9 | 10 | 11 | 12 |
|---|---|---|---|---|---|---|---|---|---|---|---|---|
| Frequency, Book A | 3 | 12 | 28 | 7 | 14 | 11 | 6 | 6 | 7 | 4 | 1 | 1 |
| Frequency, Book B | 6 | 8 | 37 | 22 | 7 | 11 | 6 | 0 | 1 | 2 | 0 | 0 |

a Draw the frequency polygons using the same axes.

b One extract was taken from a child's story and the other from an adult science fiction story.
State which is which, giving reasons for your decision.

4 The number of new designs being introduced by two furniture manufacturers was:

| | 1994 | 1995 | 1996 | 1997 | 1998 | 1999 |
|---|---|---|---|---|---|---|
| Modern Design | 8 | 7 | 8 | 5 | 2 | 1 |
| Traditional Style | 4 | 4 | 3 | 4 | 6 | 7 |

Draw the frequency polygons and comment on your results.

5 Choose two daily newspapers, one full size and the other a tabloid. Compare them by drawing frequency polygons of the number of words per sentence in one hundred sentences taken from similar sections in each newspaper.
(If you keep these results, they could be used in future work to calculate means, medians, and standard deviations.)

6 The table below shows the population of males and females in 1993 in the UK.

| Age | 0–4 | 5–15 | 16–44 | 45–59 | 60–64 | 65–79 | 80+ |
|---|---|---|---|---|---|---|---|
| Male population (thousands) | 1929 | 4276 | 12 258 | 5270 | 1355 | 3042 | 726 |
| Female population (thousands) | 1834 | 4059 | 11 840 | 5312 | 1418 | 3840 | 1643 |

(Source: *The Office of Population Censuses and Surveys*)
Draw the frequency of polygons and comment on your results.

# 18 Correlation

All the graphs and charts used so far to represent data have been concerned with only one variable, e.g. height, holiday destination, number of goals scored. In many statistical surveys, however, the data collected connect two variables, e.g. height *and* weight; number of police cameras on a stretch of motorway *and* number of speeding convictions; age *and* reaction time.

Data of this type is collected because it is believed that there is a link between the two variables.

## 18.1 Scatter diagrams

The usual method of representing these types of data is by a **scatter diagram,** also called a **scatter graph** or **scattergram**.

### EXAMPLE

A survey was carried out by a group of eight students in which the height and weight of each student was measured. The results were recorded in pairs (e.g. the student with height **164 cm** weighed **58.2 kg**).

| Height (cm) | 164 | 152 | 173 | 158 | 177 | 173 | 179 | 168 |
|---|---|---|---|---|---|---|---|---|
| Weight (kg) | 58.2 | 50.8 | 60.3 | 56.0 | 76.2 | 64.2 | 68.8 | 60.5 |

Display this data on a scatter diagram.

Two axes are drawn, one for the heights and one for the weights.
(It does not really matter which is which, but, as a general rule, the first set of data is recorded along the horizontal axis and the second set along the vertical axis.)

Each point is plotted using the paired data as the coordinates, i.e. for the student with height 164 cm and weight 58.2 kg, the coordinates are (**164, 58.2**).

The scatter diagram shows the height/weight of eight students.

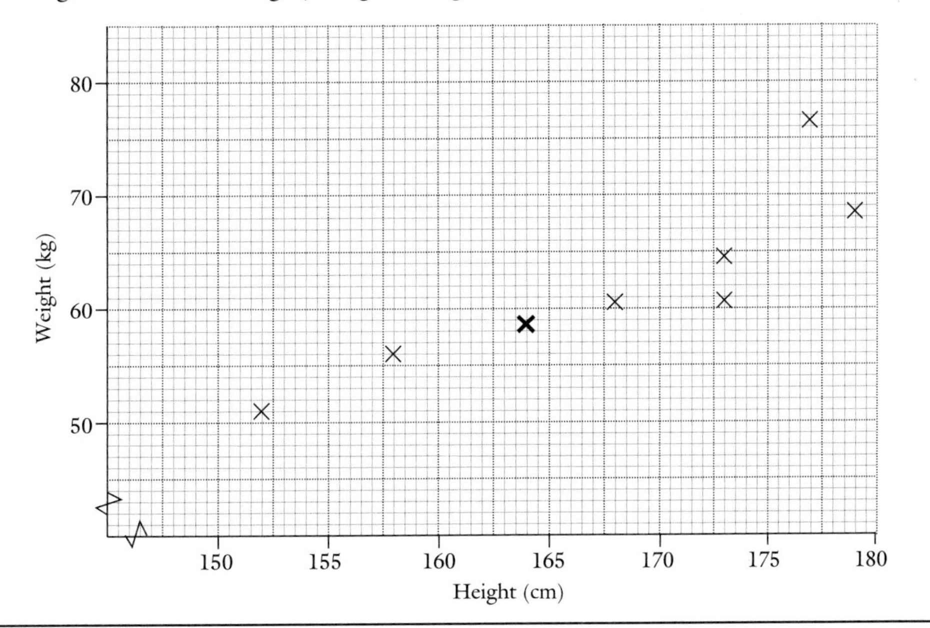

**EXERCISE 18.1**

1 Draw scatter diagrams for the following data:

a

| Mark on paper 1 | 15 | 20 | 14 | 5 | 24 | 10 | 19 | 26 | 30 |
|---|---|---|---|---|---|---|---|---|---|
| Mark on paper 2 | 22 | 34 | 50 | 20 | 66 | 32 | 50 | 34 | 60 |

b

| Time (minutes) | 10 | 15 | 20 | 25 | 30 | 35 | 40 |
|---|---|---|---|---|---|---|---|
| Weight of melting | 3.9 | 3.5 | 2.7 | 1.5 | 1.2 | 1.0 | 0.8 |

c

| Weight of parcel (kg) | 1.0 | 4.0 | 5.5 | 3.5 | 5.0 | 5.5 |
|---|---|---|---|---|---|---|
| Length of parcel (cm) | 6 | 15 | 7 | 35 | 10 | 25 |
| Weight of parcel (kg) | 4.4 | 2.2 | 2.0 | 6.0 | | |
| Length of parcel (cm) | 26 | 20 | 30 | 33 | | |

2 Draw a scatter graph for the following data.

| Shoe size | $4\frac{1}{2}$ | 10 | 5 | 9 | 7 | 6 | $5\frac{1}{2}$ |
|---|---|---|---|---|---|---|---|
| Handspan (cm) | 20.2 | 21.6 | 17.3 | 19.6 | 21.2 | 21.2 | 19.4 |
| Shoe size | 10 | 8 | 9 | 5 | 5 | 11 | |
| Handspan (cm) | 22.0 | 19.5 | 23.7 | 19.5 | 20.2 | 23.0 | |

3 The engine capacities, in cubic centimetres (cc), and the corresponding acceleration times, in seconds, for several models of car in the range of one manufacturer are given below:

| Engine size (cc) | 900 | 1000 | 1100 | 1200 | 1350 | 1500 | 1600 |
|---|---|---|---|---|---|---|---|
| Time (s) | 20.0 | 15.9 | 13.9 | 13.3 | 11.4 | 12.4 | 10.7 |
| Engine size (cc) | 1750 | 1800 | 1900 | 2000 | 1100 | 1350 | 2000 |
| Time (s) | 12.3 | 7.0 | 8.1 | 6.8 | 14.1 | 9.4 | 9.9 |

Draw a scatter graph of this data.

4 a The data given below shows the percentage of breath-tests on motorists during the Christmas/New Year period which proved positive and the number of car accidents which involved injury during the same period. The figures given are for eleven counties in England and Wales.

| % of positive tests | 9.1 | 4.1 | 8.9 | 9.2 | 5.5 | 9.2 | 7.8 |
|---|---|---|---|---|---|---|---|
| Injury accidents | 65 | 49 | 46 | 97 | 25 | 96 | 56 |
| % of positive tests | 4.2 | 5.6 | 4.9 | 10.5 | | | |
| Injury accidents | 43 | 56 | 42 | 89 | | | |

Draw a scatter graph for these eleven points.

b The same variables for another four counties are given below.

| % positive tests | 9.0 | 7.6 | 1.7 | 17.5 |
|---|---|---|---|---|
| Injury accidents | 5 | 112 | 63 | 43 |

Draw a scatter diagram for all 15 points and compare the two graphs.

## 18.2 Correlation

As might be expected, the scatter diagram in the example on page 97 shows that there is some link between height and weight but that it is not a very close one. For example, two people who are the same height may have very different weights. In general, however, the taller you are, the more you will weigh.

There are three basic types of scatter diagram:

**Diagram (i)**

The points are widely scattered. This shows that there is no link between the variables. For example, having a small head does not mean you are less intelligent than someone with a large head.

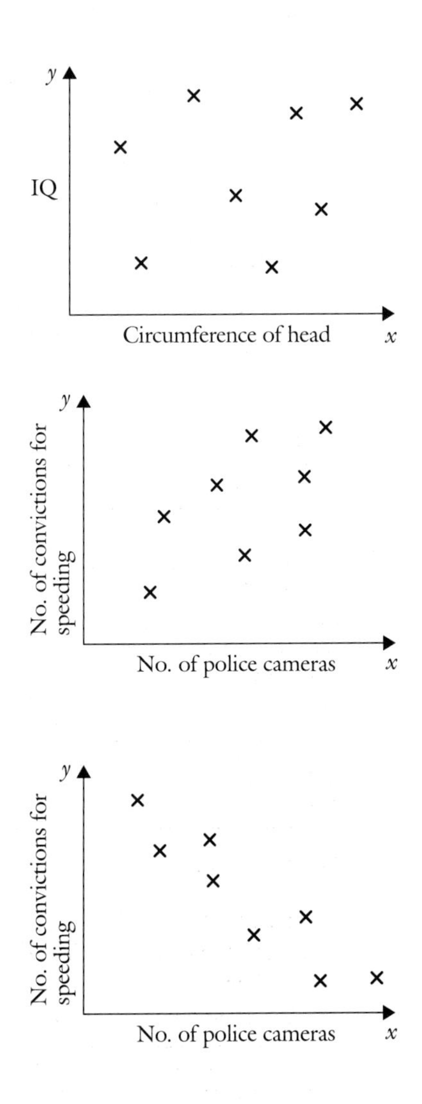


**Diagram (ii)**

This diagram might represent the number of police cameras on a motorway against the number of convictions for speeding. The points lie quite close together and show an upward trend. We might conclude that there is a weak connection, i.e. that the more cameras there are, the greater the number of motorists caught speeding.

However, the purpose of the cameras is not so much to catch motorists speeding as to deter them from speeding.

**Diagram (iii)**

This could represent the same variables after the cameras have been in place for a while and motorists are aware of them. There is still a link, but now the trend is downward, and we might conclude that the more cameras that are known to be operating, the less likely motorists are to drive fast and be caught speeding.

The closer a set of points lies to a straight line, the stronger the relationship between the variables, or, in statistical terms, the higher the **degree of correlation**.

**Diagram (i) indicates that there is no correlation**
**Diagram (ii) indicates quite a weak positive correlation**
**Diagram (iii) indicates quite a high negative correlation.**

### *EXERCISE 18.2*

**1–4** For each of the questions in Exercise 18.1, state the type of correlation (none, positive, negative) and give some indication of the degree of correlation.

**5** For each of the following sets of data, state whether there is no correlation, positive correlation or negative correlation. (It is not necessary to draw the scatter diagrams.)

a

| Day | 1 | 2 | 3 | 4 | 5 | 6 |
|---|---|---|---|---|---|---|
| No. of ripe courgettes | 3 | 5 | 4 | 7 | 9 | 9 |

b

| Height (cm) | 166 | 156 | 152 | 160 | 159 | 158 | 163 |
|---|---|---|---|---|---|---|---|
| Age (years) | 20 | 31 | 25 | 28 | 22 | 32 | 32 |

c

| % seeds germinating in compost 1 | 51 | 60 | 32 | 70 | 20 | 78 |
|---|---|---|---|---|---|---|
| % seeds germinating in compost 2 | 55 | 75 | 70 | 58 | 82 | 50 |

d

| W.p.m. typing test part 1 | 56 | 43 | 49 | 31 | 58 | 60 | 35 | 35 | 44 | 38 |
|---|---|---|---|---|---|---|---|---|---|---|
| W.p.m. typing test part 2 | 58 | 44 | 47 | 37 | 57 | 57 | 30 | 31 | 44 | 43 |

6 The following table shows the amount of cod landed in the years 1981–86 and percentages of babies vaccinated against measles for the same years.

| Cod landed (1000 tonnes) | 116 | 114 | 112 | 91 | 90 | 77 |
|---|---|---|---|---|---|---|
| % vaccinated against measles | 54 | 56 | 59 | 63 | 68 | 71 |

a Draw the scatter graph for the above data.

b Can we deduce from this graph that the greater the number of babies vaccinated, the smaller the cod catch will be?

c Explain the high degree of correlation.

It is important to realise that correlation between two sets of data does not mean that the changes in one variable **cause** the changes in the other variable. In other words, being a particular weight does not **cause** you to be a particular height, nor does being a particular height decide what weight you will be.

Although height is linked to weight, there are many other factors which determine your weight.

There is a story that a survey carried out in Sweden paired the number of storks nesting on houses and the number of babies born in the area. The results showed a positive correlation.

We cannot conclude from this that a large number of storks in the area causes a large number of babies to be born, or vice versa.

The reason for the seeming connection is likely to be a third factor, the number of new houses. Storks prefer to nest on new chimneys and new housing estates attract young couples starting families.

When analysing scatter graphs, care should be taken not to jump to conclusions.

It is possible to find a high degree of correlation between variables which are very unlikely to be connected (particularly if only a few points are plotted).

## 18.3 Line of best fit

If there is a perfect (linear) correlation between two variables, the points plotted will lie on a straight line.

This means that one variable is proportional to the other and an equation of the form $y = mx + c$ can be found to fit the line through the points.

When experimental data is collected, however, there is usually some error in the readings and so the points plotted will not lie exactly along a straight line but close to one.

### EXAMPLE 1

An experiment was carried out to measure the height to which a rubber ball bounced after being dropped from various heights.

| Height of drop (m) | 0.50 | 0.75 | 1.00 | 1.25 | 1.50 | 1.75 | 2.00 |
|---|---|---|---|---|---|---|---|
| Height of bounce (m) | 20 | 25 | 50 | 60 | 75 | 75 | 100 |

Scatter graph:

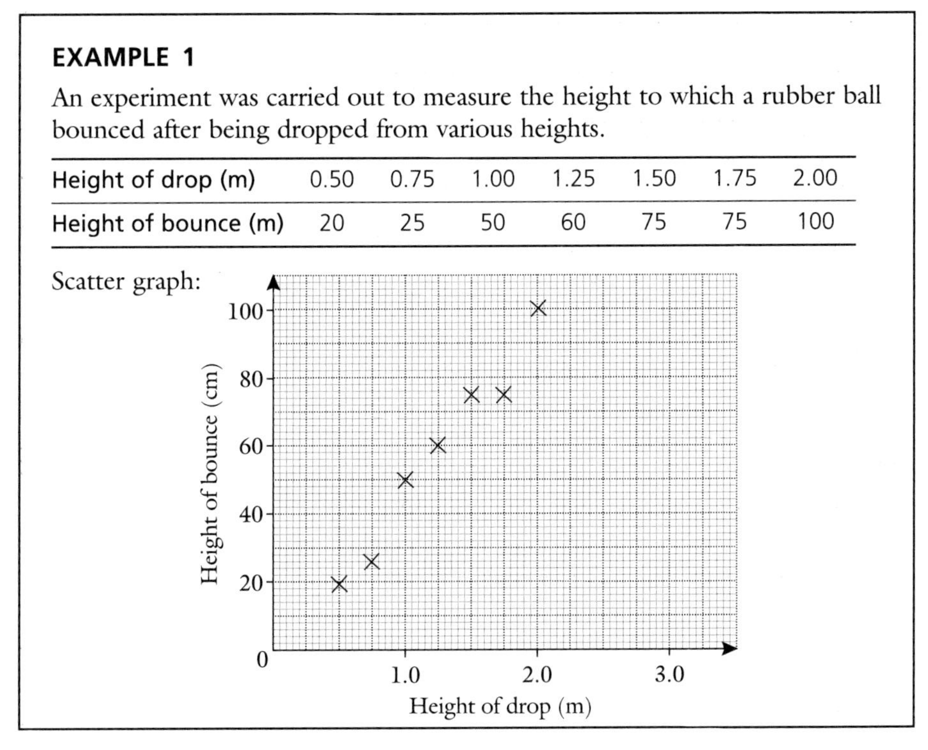


The scatter graph shows that there is a close linear relationship between height of drop and height of bounce. It is not possible to draw a straight line through all the parts so the line which **best fits** all the points is drawn.

All lines of best fit pass through the point $(\bar{x}, \bar{y})$ where $\bar{x}$ and $\bar{y}$ are the means of the first and second sets of data respectively. It is not necessary, for FSMU at Foundation or Intermediate level, to find these means before drawing the line of best fit unless asked to do so.

### EXAMPLE 2

Draw the line of best fit for the data of Example 1 and use the line to predict the height of bounce when the drop is 1.85 metres.

Draw, by eye, the line which looks to be the 'line of best fit'.

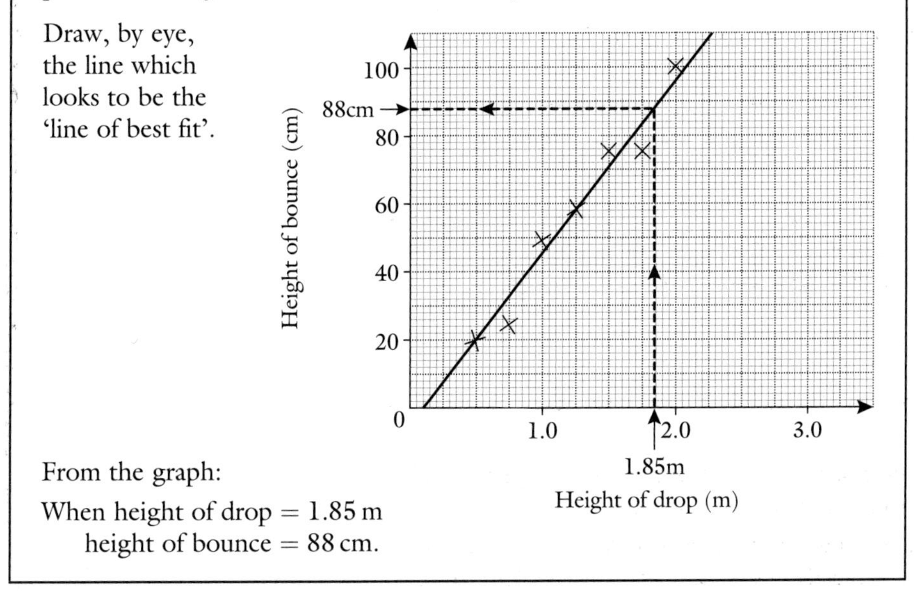


From the graph:

When height of drop = 1.85 m
height of bounce = 88 cm.

## EXERCISE 18.3

1 **a** Draw the scatter diagram for the following data or use your answer to question **1 b** of Exercise 18.1.

| Time (minutes) | 10 | 15 | 20 | 25 | 30 | 35 | 40 |
|---|---|---|---|---|---|---|---|
| Weight of melting ice block (kg) | 3.9 | 3.5 | 2.7 | 1.5 | 1.2 | 1.0 | 0.8 |

**b** Calculate $\bar{x}$ and $\bar{y}$.

**c** Draw the line of best fit on your scatter graph.

**d** From the graph, find the values of the time when the weight is:
(i) 2.0 kg (ii) 3.0 kg

2 **a** The graph below shows the mean weight for a man of medium build and given height.

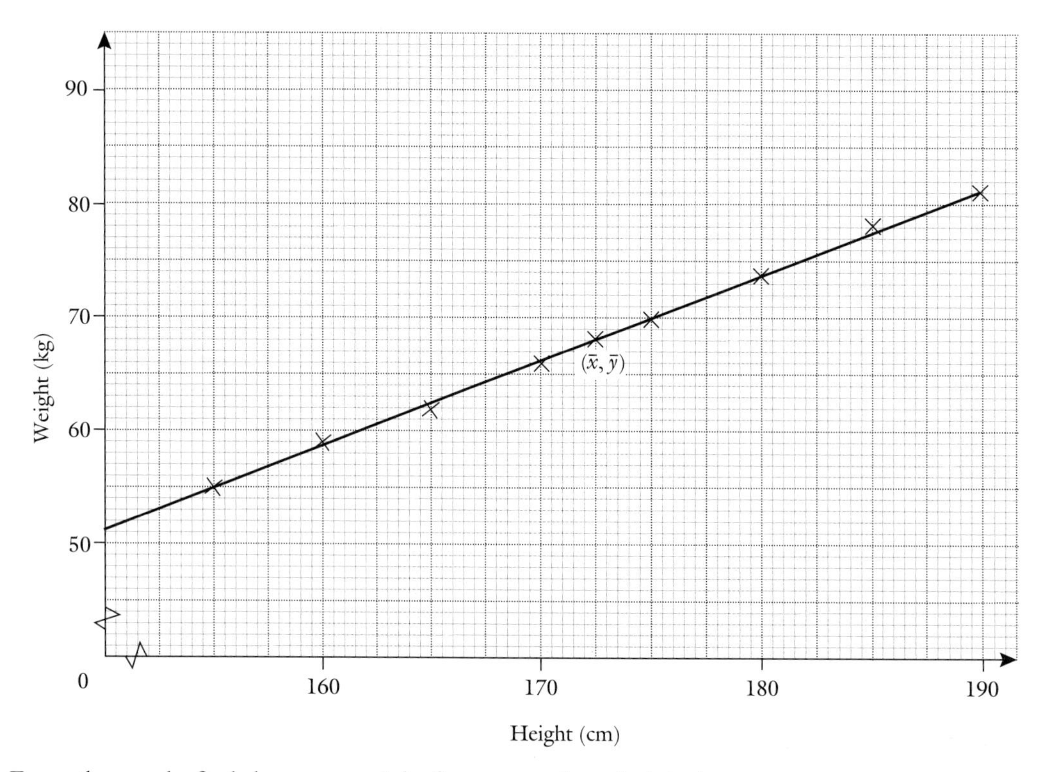


From the graph, find the mean weight for a man whose height is:
(i) 167.5 cm (ii) 180 cm

**b** The corresponding values for a woman's mean weight are:

| Height (cm) | 142 | 147 | 152 | 157 | 162 | 167 | 172 | 177 |
|---|---|---|---|---|---|---|---|---|
| Weight (kg) | 46 | 49 | 51 | 54 | 58 | 62 | 65 | 69 |

(i) Plot the scatter graph.
(ii) Draw the line of best fit.
(iii) Find the mean weight for a woman of height 165 cm.

# 19 Probability

## 19.1 Introduction

**Probability** is a measure of how likely something is to occur. What occurs is an **outcome**.

This likelihood or probability can be represented on a sliding scale from the probability when an event is certain to occur to the probability when the event cannot occur.

Suppose twenty children in class 2X are studied. It is found that 19 do not wear glasses, and 1 does wear glasses. 11 of the 20 children are girls and 9 are boys.

Suppose a child is picked at random. The likelihood of the following events occurring fits into the following pattern.

| Event | Likelihood |
|---|---|
| The child studied is in class 2X | Certainty |
| The child does not wear glasses | High probability |
| The child is a girl | Just over half |
| The child is a boy | Just under half |
| The child wears glasses | Highly unlikely |
| The child's age is over 70 | Impossible |

Since 19 out of 20 children do not wear glasses, the likelihood that a child selected at random does not wear glasses is 19 out of 20, or $\frac{19}{20}$. Therefore, the probability that a child selected at random is not wearing glasses is $\frac{19}{20}$.

If all 20 children are wearing shoes, the likelihood that a child selected at random is wearing shoes is 20 out of 20, i.e. $\frac{20}{20}$, which is 1. Therefore, the probability is 1.

None of the children have green hair. The probability that a child selected at random has green hair is 0 out of 20, i.e. $\frac{0}{20}$, which is 0.

**If an event is bound to occur, then its probability is 1.**

**If an event cannot occur, then its probability is 0.**

The probability of a child not wearing glasses is $\frac{19}{20}$.

The probability of a child wearing glasses is $\frac{1}{20}$.

The probability of a child wearing glasses or not wearing glasses is 1 (i.e. a certainty).

Note that $\frac{19}{20}$ (not glasses) + $\frac{1}{20}$ (glasses) = $\frac{20}{20}$ = 1 (certainty).

From this we can work out that:

(i) **The total probability for all possible outcomes is 1.**

(ii) **Probability of event happening = 1 − Probability of it not happening**

**The events in the initial example occur with the following probabilities:**

| Event | Likelihood | Probability |
|---|---|---|
| The child studied is in class 2X | Certainty | 1 |
| The child does not wear glasses | High probability | $\frac{19}{20}$ |
| The child is a girl | Just over half | $\frac{11}{20}$ |
| The child is a boy | Just under half | $\frac{9}{20}$ |
| The child wears glasses | Highly unlikely | $\frac{1}{20}$ |
| The child's age is over 70 | Impossible | 0 |

## 19.2 Probability from theory and experiment

Probability can be found either by theory or by experiment.

The theory method relies on logical thought; the experimental method relies on the result of repetition of the event producing results which are taken to be typical.

**EXAMPLE**

Find the probability of getting a head when tossing an unbiased coin.

*Theory method*
'Unbiased' means both events (head, tail) are equally likely to occur.

Heads and tails are the only outcomes,
$\therefore$ Probability of head + Probability of tail = 1
(since total of all possible outcomes is 1)

Probability of head = Probability of tail
$\therefore$ Probability of tail $= \frac{1}{2}$.

*Experimental method*
Toss an unbiased coin 100 times, and count the number of tails you obtain. You might obtain 49 tails.
$\therefore$ The experimental probability of getting a tail is $\frac{49}{100}$.

If the event is repeated many times, it is usual for the experimental probability to be close to the theoretical probability, but not to be exactly the same.

For the probability of a person being right-handed, there is no theory to use. This would be found by experiment. For example, ask 100 people and see how many are right-handed. The more people you ask, the more likely it is that your probability is accurate, but it must be appreciated that the probabilities found by experiment are not exact. The result may also be different if the experiment is repeated with another group.

Similarly, if the probability of a car being red is $\frac{1}{4}$, the expected number of red cars in a car park containing 80 cars is

$$\frac{1}{4} \times 80 = 20$$

## EXERCISE 19.1

**1** The probability that a street light is lit is $\frac{3}{8}$.
What is the probability that it is not lit?

**2** You toss a coin, marked with a head on both sides.
What is the probability of getting a tail?

**3** The probability of a plane being late is $\frac{2}{3}$.
What is the probability of a plane not being late?

**4** Jane has a bag containing 6 oranges and 4 apples. She selects a fruit from the bag at random.
What is the probability that it is a pear?

**5** Probabilities can be estimated **either** by making a subjective estimate **or** by making use of statistical evidence.
State the method which would be used in the following cases.

**a** The probability that more women than men will read the news on television.

**b** The probability that man will live on the moon in the year 2050.

**c** The probability that the next vehicle passing a college will be a motor cycle.

**d** The probability that the next book issued at a library will be a novel.

**e** The probability that there will be a cure for deafness within 10 years.

(SEG S95)

**6** In an experiment a drawing pin is thrown. The number of times it lands point up is recorded.

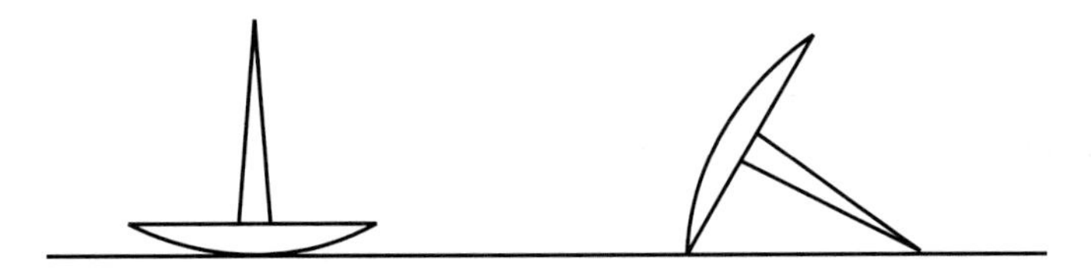

The drawing pin is thrown 10 times and lands point up 6 times.

**a** From these data estimate the probability that it lands point up.

The drawing pin is thrown 100 times and lands point up 57 times.

**b** From these data estimate the probability that it lands point up.

**c** Give a reason why the second answer is a more reliable estimate.

(SEG S94)

# 20 Algebra

## 20.1 The basics of algebra

### From words to symbols

Every Saturday the milkman delivers the following to 16 Cromarty Street:
3 loaves of bread, 4 pints of milk and a dozen eggs.

He then presents his bill.

A sentences such as 'the cost of three loaves of bread, four pints of milk and a dozen eggs' is difficult to deal with mathematically.

Algebra is a mathematical language which enables us to deal with problems more easily.

Suppose we let:

$b$ stand for 'the cost of a loaf of bread'
$m$ stand for 'the cost of a pint of milk'
and $e$ stand for 'the cost of a dozen eggs'.

The milkman's bill for 3 loaves of bread, 4 pints of milk and 1 dozen eggs becomes

$$3b + 4m + 1e \text{ or } 3b + 4m + e$$

where $3b$ means 3 times $b$ (or $3 \times b$).
We can miss out the 1 in front of the $e$
($e$ means $1 \times e$).

The foundation of algebra, as a science of equations, was laid down in the year 825 by an Islamic mathematician, Muhammad Ibn Musa al-Khwarizmi.

**EXAMPLE 1**

Don thinks of a number, multiplies it by 3, adds 4 and then divides this answer by the original number.
Write his final number in algebraic terms.

Let his first number be $n$

Multiply by 3: $3n$

Add 4: $3n + 4$

Divide by the original number: $\dfrac{3n+4}{n}$

Final number $= \dfrac{3n+4}{n}$

**EXAMPLE 2**

A holiday has two prices, one for adults and the other for children.
Write down an algebraic expression for the cost of a holiday for two adults and three children.

Let the cost for one adult be £$A$ and the cost for one child be £$C$:

Cost for 2 adults $=$ £$2A$

Cost for 3 children $=$ £$3C$

Cost of the holiday $=$ £$2A$ $+$ £$3C$ $=$ £$(2A + 3C)$

**EXERCISE 20.1**

Simplify the following phrases and sentences by translating them into algebraic expressions:

**1** The sum of seven times $x$ and five times $y$.

**2** The difference between $q$ and twice $p$.

**3** Double the number $a$ and divide by the number $b$.

**4** A number is formed by multiplying a number $x$ by five and then subtracting eight.

**5** A number is formed by dividing a number $y$ by four and then adding six.

**6** The cost of admission to an art exhibition for 4 adults and 3 children.

**7** To make purple paint, an artist mixes 4 cl of blue paint, 6 cl of red paint and 2 cl of white paint.

**8** The cost of 4 reams of paper, 3 packs of pens and 2 bottles of Tipp-ex®.

**9** The charge levied by a tax adviser who on your behalf writes 6 letters, makes 8 phone calls and has 2 interviews with you.

**10** The number of beds in a hospital that has 22 NHS wards and 3 private wards.

**11** The time spent by a social worker whose case-load is 4 teenagers, 5 retired couples and 3 single mothers.

**12** The entrance fee to a leisure centre for 3 adults and 5 children.

**13** The total paid in deposits to a travel agent who has taken 15 bookings in a day: 7 for a flight package, 5 for a ferry crossing to France, and the remainder for holidays in England.

**14** The cost of fitting a door in a house for which a carpenter needs 2 handles, 3 hinges, and 22 screws.

**15** Owen's commission if he sells 55 appliances of which 28 are fridges, 12 are microwaves, and the rest are washing machines.

## Collecting like terms

Mrs Goodman empties her son's pockets on washing day and finds two sweets, one dirty hanky, two pencils, two more sweets and another pencil.

The contents of the pocket can be listed in algebraic terms as

$$2s + 1h + 2p + 2s + 1p$$

The list can be simplified if articles of the same kind are added together. This is called **collecting like terms**. If we collect the like terms from Master Goodman's pocket, the list becomes $h + 3p + 4s$ (which cannot be simplified further as each term is different).

## Addition and subtraction

**Only like terms may be added or subtracted.**

Like terms are those which are multiples of the same algebraic variable.

For example, $3a$, $7a$ and $-8a$ are all like terms.

The expression $3a + 7a - 8a$ can be simplified to $2a$.

The terms $7a$, $3b$ and $4c$ are unlike terms and the expression $7a + 3b + 4c$ cannot be simplified any further.

**EXAMPLE 1**

Simplify the following expression:

$$2a + 3c + a + 4b + 2c + b$$

$$\begin{aligned} 2a + 3c + a + 4b + 2c + b &= (2a + a) + (4b + b) + (3c + 2c) \\ &= 3a + 5b + 5c \end{aligned}$$

**EXAMPLE 2**

Simplify the following by collecting like terms:

$$5x + 2y - 4z + 3x - y + z$$

$$\begin{aligned} 5x + 2y - 4z + 3x - y + z &= (5x + 3x) + (2y - y) + (z - 4z) \\ &= 8x + y - 3z \end{aligned}$$

*EXERCISE 20.2*

Simplify the following by collecting terms:

1 $2x + 3x$

2 $8n - n$

3 $3a - a + 6a$

4 $5y - 6y + 3y$

5 $3a + 2a + c + b + 2a + 3c + 5b$

6 $x + 2y + z + 3y + 8z$

7 $5p + 3q + 2r - q + 7p - 3p - r$

8 $2x - y + 3x - 2y + z$

9 $a + b - 2a + c - 2b + 3c$

10 $3x - y + \frac{1}{2}y - 2\frac{1}{2}y + 8y$

## Substitution

The letters in an algebraic expression stand for numbers or amounts.

For example, in the expression $3b + 4m + e$, $b$ stands for the cost of a loaf of bread, $m$ for the cost of a pint of milk and $e$ for the cost of a dozen eggs.

The advantage of the letters is that the amounts are not fixed, they can change or vary.

The letters are called **variables**. The numbers, which are constant, are called **coefficients**.

If a loaf of bread costs 54p, a pint of milk 39p and a box of eggs 63p (i.e. $b = 54$, $m = 39$, $e = 63$), then

$$\text{Weekly bill} = 3b + 4m + e$$
$$= (3 \times 54) + (4 \times 39) + 63$$
$$= 381 \text{ pence}$$

If, however, prices rise and $b = 56$, $m = 41$, $e = 66$, then

$$\text{Weekly bill} = 3b + 4m + e$$
$$= (3 \times 56) + (4 \times 41) + 66$$
$$= 398 \text{ pence}$$

**EXAMPLE 1**

Evaluate $\frac{3x - y}{x + y}$ when $x = 3$, $y = -1$

$$\frac{3x - y}{x + y} = \frac{(3 \times 3) - (-1)}{3 + (-1)} = \frac{9 + 1}{3 - 1} = \frac{10}{2} = 5$$

*EXERCISE 20.3*

1 Evaluate the following expressions when $x = 3$, $y = 5$, $z = 2$.

a $2x + y$
b $3x - 2y$
c $y + 4z$
d $\frac{12}{z}$
e $\frac{y}{10}$
f $4z - 6$
g $3z - 12x$
h $2x - 5z$
i $x + y - 4z$
j $xy + z$
k $y - xz$
l $yz - 4x$

2 Evaluate each of the expressions in question 1 when $x = -1$, $y = 3$, $z = -2$.

3 Evaluate the following expressions if $x = 4$, $y = 3$, $z = -2$:

a $y^2$
b $xyz$
c $2z^2$
d $\frac{x}{y}$
e $xz + y$
f $\frac{y}{2z}$
g $2x^2 - yz$
h $\frac{x + y}{x - y}$
i $x(x + y)$
j $\frac{2y}{x - z}$
k $\frac{x + 2z}{y}$
l $(x + y)(x + z)$

4 Evaluate each of the expressions in question 3 when $x = -2$, $y = \frac{1}{2}$, $z = -1$.

## Multiplication and division

Multiplication of algebraic terms is usually easier than multiplication in arithmetic. For example:

$a$ multiplied by $4 = a \times 4 = 4a$

$p$ multiplied by $q = p \times q = pq$

$x$ divided by $y = x \div y = \frac{x}{y}$

$a$ multiplied by $a = a \times a = a^2$

Algebraic terms are usually listed in alphabetical order with the constant first.

The rules for directed numbers are the same as on pp. 6–7. For example:

$$(+x) \times (+y) = xy$$

$$(-x) \times (+y) = -xy$$

$$(+x) \div (-y) = -\frac{x}{y}$$

$$(-x) \div (-y) = \frac{x}{y}$$

**EXAMPLE 1**

Simplify $5 \times p \times 6 \times q \div 3 \div r$

Collect numbers and letters separately:

$$5 \times p \times 6 \times q \div 3 \div r = (5 \times 6 \div 3) \times (p \times q \div r)$$
$$= \frac{10pq}{r}$$

**EXAMPLE 2**

Simplify $(3x) \times (-2y) \times (-z)$

$$(3x) \times (-2y) \times (-z) = (3 \times -2 \times -1) \times (x \times y \times z)$$
$$= 6xyz$$

**EXAMPLE 3**

Simplify $3a \times (2b)^2 \times (-a^2)$

$$3a \times (2b)^2 \times (-a^2) = (3 \times 2^2 \times -1) \times a \times b^2 \times a^2$$
$$= -12a^3b^2$$

**EXERCISE 20.4**

Simplify:

1 $5 \times x$

2 $3 \times m \times n$

3 $2 \times y \div z$

4 $a \times c \times b$

5 $2 \times p \times 3 \times r \times q$

6 $c \div 2d \times 8b$

7 $(-2x) \times (-3y)$

8 $(4p) \div (-2q)$

9 $(6x) \times (-3y) \div (-2z)$

10 $a \times a \times a$

11 $a \times a \times b \times b$

12 $3a^2 \times a^2$

13 $3a \times b^2 \times a^2$

14 $2a^2b \times -3bc$

15 $3ab^2 \times 2ab \times bc^2$

16 $(-3a)^2 \times b$

17 $(-2ab) \times (3b)^2$

18 $(-2a) \times (-b) \div (-b)$

19 $x^2y \times xy^2$

20 $12ab \div (-4c^2) \times (3ac)$

## 20.2 Indices

$$a \times a \times a = a^3$$

The number 3 is called an **index**. It shows, or indicates, the number of *as* which have been multiplied together to give the third power of *a*.

$$a^1 \times a^2 = a \times a \times a \qquad = a^3 \quad \text{i.e.} \quad a^{(1+2)}$$

$$a^3 \times a^2 = a \times a \times a \times a \times a = a^5 \quad \text{i.e.} \quad a^{(3+2)}$$

In general: $\boldsymbol{a^x \times a^y = a^{(x+y)}}$

$$a^3 \div a^2 = \frac{a \times a \times a}{a \times a} = a \quad \text{i.e.} \quad a^{(3-2)}$$

$$a^5 \div a^3 = \frac{a \times a \times a \times a \times a}{a \times a \times a} = a \times a = a^2 \quad \text{i.e.} \quad a^{(5-3)}$$

In general: $\boldsymbol{a^x \div a^y = a^{x-y}}$

$$a^3 \div a^5 = \frac{a \times a \times a}{a \times a \times a \times a \times a} = \frac{1}{a \times a} = \frac{1}{a^2}$$

but $a^3 \div a^5 = a^{(3-5)} = a^{-2}$ i.e. $\frac{1}{a^2} = a^{-2}$

In general: $\boldsymbol{a^{-x} = \frac{1}{a^x}}$

$$a^3 \div a^3 = \frac{a \times a \times a}{a \times a \times a} = 1$$

but $a^3 \div a^3 = a^{(3-3)} = a^0$

In general: $\mathbf{a^0 = 1}$

The rules for indices can be summarised as follows:

$$\boldsymbol{a^x \times a^y = a^{(x+y)}}$$

$$\boldsymbol{a^x \div a^y = a^{(x-y)}}$$

$$\boldsymbol{a^{-x} = \frac{1}{a^x}}$$

$$\boldsymbol{a^0 = 1}$$

**EXAMPLE**

Simplify $x^5 \div x^3 \times x^2$

$$x^5 \div x^3 \times x^2 = \frac{x \times x \times x \times x \times x \times x \times x}{x \times x \times x} = x^4$$

or $x^5 \div x^3 \times x^2 = x^{(5-3+2)} = x^4$

**EXERCISE 20.5**

Simplify the following expressions:

1 $x^5 \times x^3$

2 $x^3 \times x \times x^2$

3 $x^7 \div x^4$

4 $x4 \div x^5$

5 $x^4 \div x^4$

6 $x^{-2}$

7 $x^0$

8 $3x^3 \times 2x \div 4x^2$

## 3 Formulae

### ubstitution in formulae

A second important use of algebra is to enable a relationship between two or more quantities to be expressed in a short but easily understood form. This is called a **formula**.

The volume of a cylinder is found by multiplying the area of the circular cross-section of the cylinder by its height. This is more neatly expressed by the formula

$$V = \pi r^2 h$$

In order to calculate the volume, the given values of $r$ (radius) and $h$ (height) are substituted into the formula.

**EXAMPLE**

Ian invested £5000 for 4 years and his investment had grown to £6241.

Using the formula $R = \sqrt[n]{\frac{A}{P}} - 1$

where $A$ is the final amount,
$P$ is the Principal, or amount invested,
and $n$ is the number of years,

find $R$, the annual rate of return on Ian's investment.

Substituting $A = 6241$, $P = 5000$ and $n = 4$

into $R = \sqrt[n]{\frac{A}{P}} - 1$ gives

$$R = \sqrt[4]{\frac{6241}{5000}} - 1$$

$$= \sqrt[4]{1.2482} - 1$$

To find $\sqrt[4]{1.2482}$ use the calculator keys 1.2482, INV, $x^y$, 4, as described on page 13.

Hence $R = 1.05699 - 1$

$$= 0.05699$$

The annual percentage rate is 5.7%.

**EXERCISE 20.6**

1 The formula for the amount £ $A$ accruing when a principal £ $P$ earns compound interest at an annual rate of $R$% for $n$ years is given by the formula

$$A = P\left(1 + \frac{R}{100}\right)^n$$

Use this formula to find the amount accruing when:

**a** £800 is invested at 5% for 4 years

**b** £6000 is invested at 4% for 7 years

**c** £2100 is invested at 7% for 5 years.

**2** Using the formula $R = \sqrt[n]{\frac{A}{P}} - 1$, which was given in the example on page 112, find the annual percentage rate when:

**a** a principal of £7000 is increased to £9000 in 5 years,

**b** a principal of £7500 is increased to £9000 in 8 years,

**c** a principal of £4000 is increased to £7000 in 3 years,

**d** a principal of £1700 is increased to £3900 in 6 years,

**3** Helen sells $n$ home-knitted scarves at a profit of £$p$ each. This price includes VAT at $V$%.

The VAT Sarah pays is given by £$\frac{npV}{100 + V}$.

**a** Helen sells 40 scarves at a profit of £17 each. The rate of VAT is $17\frac{1}{2}$%. How much VAT does Helen have to pay?

**b** Sue sells a different scarf at a profit of £21 each. The rate of VAT is $17\frac{1}{2}$%. On these scarves Sue pays £147 in VAT. How many of these scarves does Sue sell?

**4** The cost, £$C$, of transporting a yacht from a manufacturer to a purchaser is estimated to be given by $C = 80 + 0.76m + 0.2s$, where $m$ miles is the distance to be travelled by road and $s$ miles is the distance to be travelled by sea. The purchaser wants the yacht delivered to Cherbourg. Delivery will involve a road distance of 50 miles and a sea distance of 120 miles.

Find the cost.

**5** The cost, £$C$, of printing $n$ posters for a concert is given by $C = 5.2 + 0.11n$.

**a** Find the cost of printing:

(i) 100 posters

(ii) 1000 posters.

**b** Find the cost per poster when 400 are printed.

**6**

A wedding dress is to be finished with an embroidered front bodice, decorated with pearls. The cost £$C$ of the embroidered section is given by $C = h(2b + 3a)/20$, where $h$ cm is the depth of the embroidered panel, $a$ cm is the minimum width and $b$ cm is the maximum width, as shown in the diagram.

Find the cost of the embroidery when

**a** $h = 25$, $a = 20$ and $b = 25$

**b** $h = 18$, $a = 18$ and $b = 19$.

# 21 Graphs

## 21.1 Graphs and curves

This unit deals with line graphs. The lines may be straight or curved, but they illustrate a relationship between two quantities. This relationship must be clear to anyone looking at the graph.

Graphs are used by many bodies to convey information quickly and with impact!

For example, these graphs were used by:

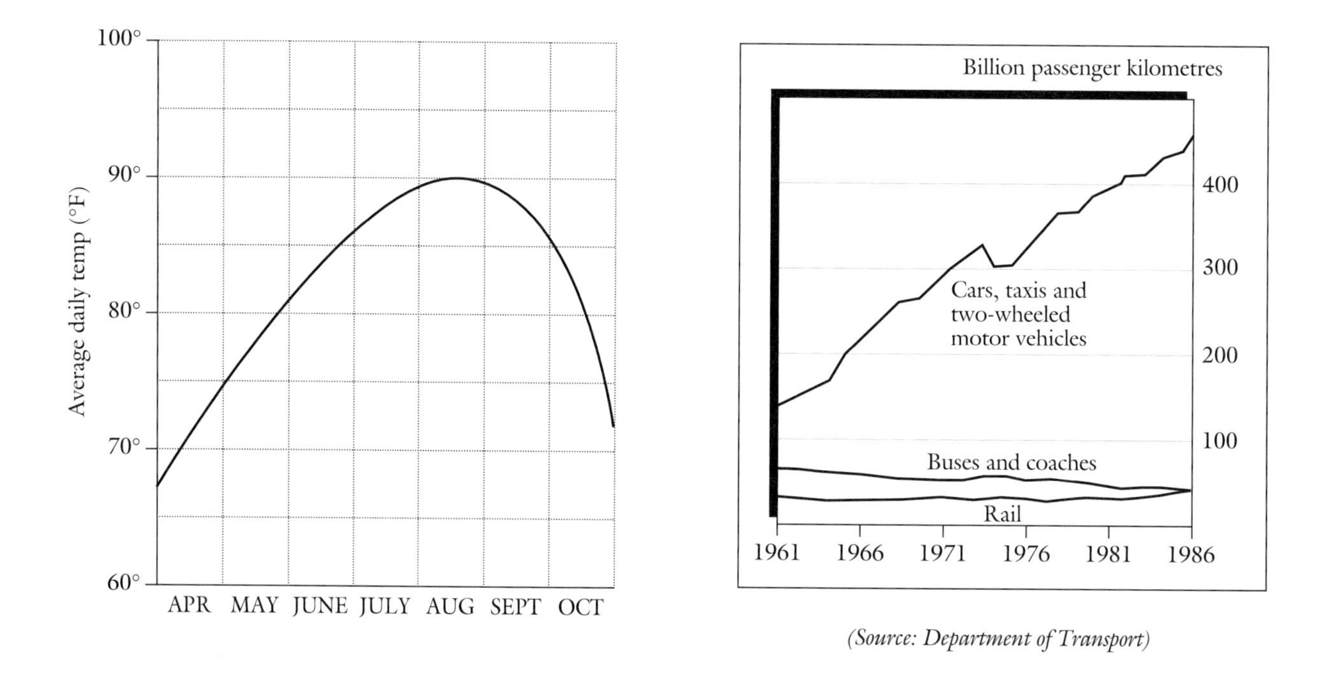


*(Source: Department of Transport)*

Many graphs involve time, which is always measured along the horizontal axis. The vertical axis then provides a measure of something that changes with time.

Whenever one measurable quantity changes as a result of another quantity changing it will be possible to draw a graph.

## 21.2 The interpretation of graphs

Once a graph has been drawn, information can be found from it very quickly.

### EXAMPLE 1

The graph shows a cross-section through a river bed. The distance from point A on one bank of the river to point B, directly opposite on the other bank is 20 m.

Soundings were taken of the depth of the water at various points and plotted against the distance from A.

Find, from the graph:

**a** the depth of the river at a distance of 12 m from A

**b** the distances from A at which the depth of the river is 2.8 m.

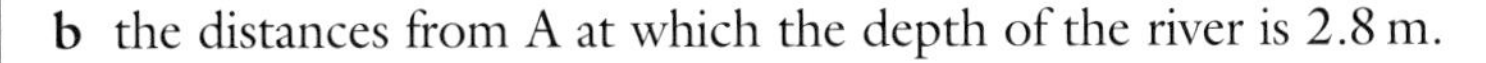

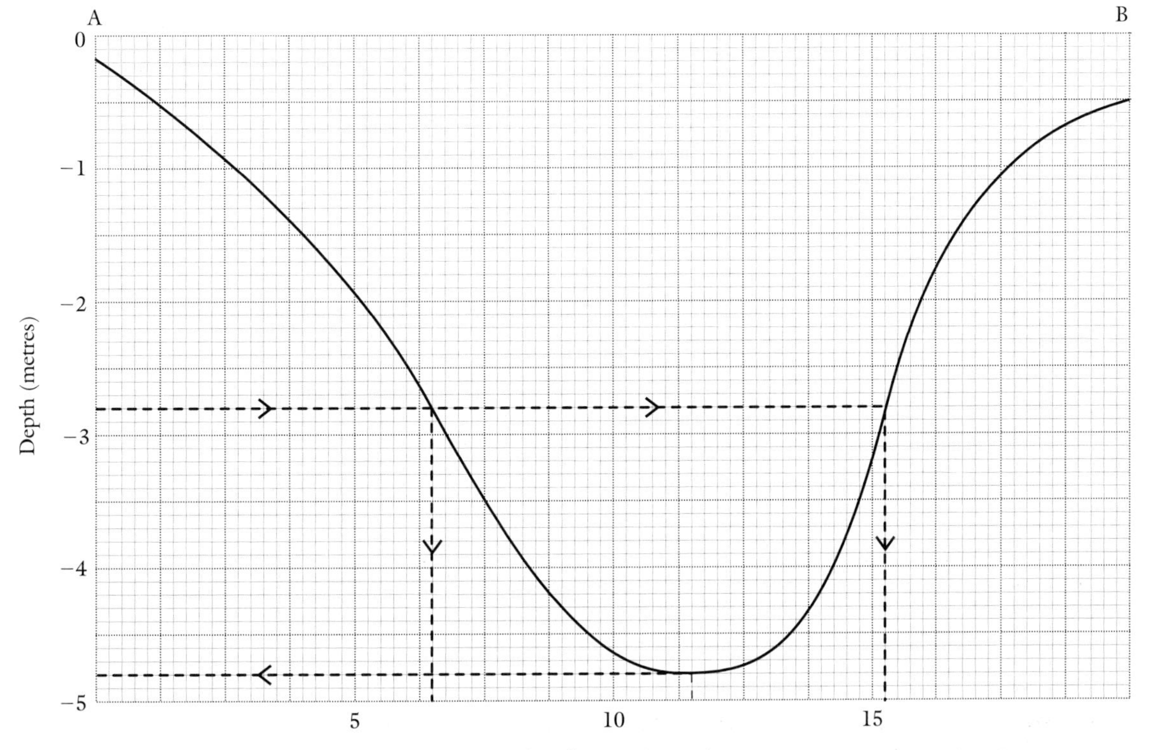


**a** The distance 12 m is located on the horizontal axis and a vertical line is drawn from this point to meet the graph. A horizontal line is then drawn across to the vertical axis and the distance is read off.
The distance is -4.8 m, i.e. the depth of the water is 4.8 m.

**b** The depth of 2.8 metres is located on the vertical axis at the point −2.8 and a horizontal line is drawn from this point which meets the curve at two points.
Vertical lines are drawn down to the horizontal axis and the two distances are read off.

The distances from A are 6.50 m and 15.25 m.

If a connection between two variables can be expressed algebraically, i.e. by an equation, then every point lies on the line or curve. If one value is known, the other can be found exactly, or as accurately as the scale of the graph allows.

For example, using the graph which shows the depth of a river (Example 1 above) the depth of the river can be given correct to the nearest 10 cm.

Often, however, the connection is not exact (as in Example 2 below) and, given one variable, only a rough estimate of the other variable can be found.

### EXAMPLE 2

Each year thousands of tourists visit Britain, and the money which they spend here is a very welcome addition to the British economy. The larger the number of tourists staying in Britain, the larger the amount of money spent will be. However, the relationship between the number of tourists and the amount of money spent is not an exact linear relationship. For example, double the number of tourists will not necessarily spend double the amount of money.

The line drawn in the graph below is the one which best represents the relationship between the number of tourists and the money they spend.

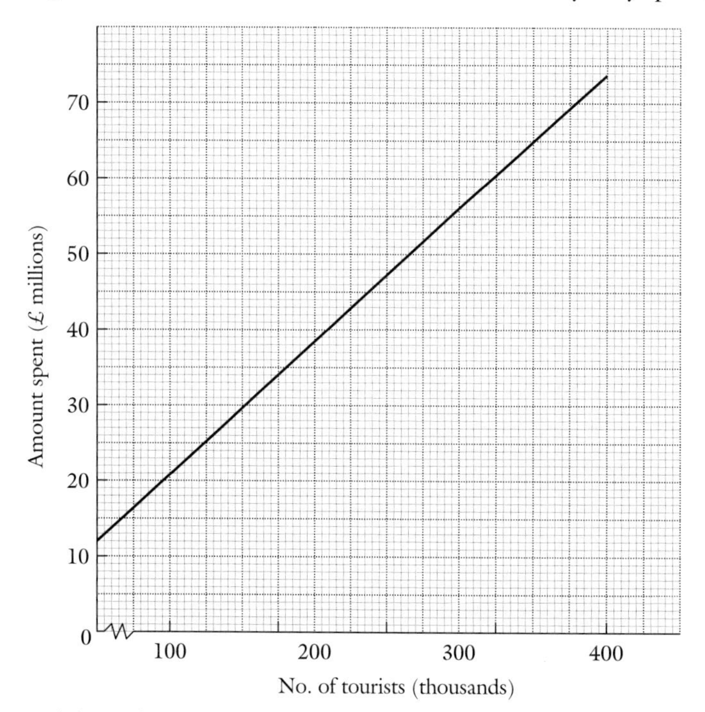


Values read from the graph are likely to be approximate, but the graph could be used to make predictions.

**a** Approximately how much money is spent by 250 thousand tourists?

**b** Predict the number of tourists needed to produce an income of £60 million.

From the graph:

**a** The amount of money spent is £47.5 million.

**b** The number of tourists required is 320 000.

## EXERCISE 21.1

In the following questions state whether the answer found from the graph is exact, within the limits of the graph's accuracy, or whether it is a rough estimate.

1 The graph shows the relationship between the percentage of rejected mugs found by different potteries and the selling price of the mug.

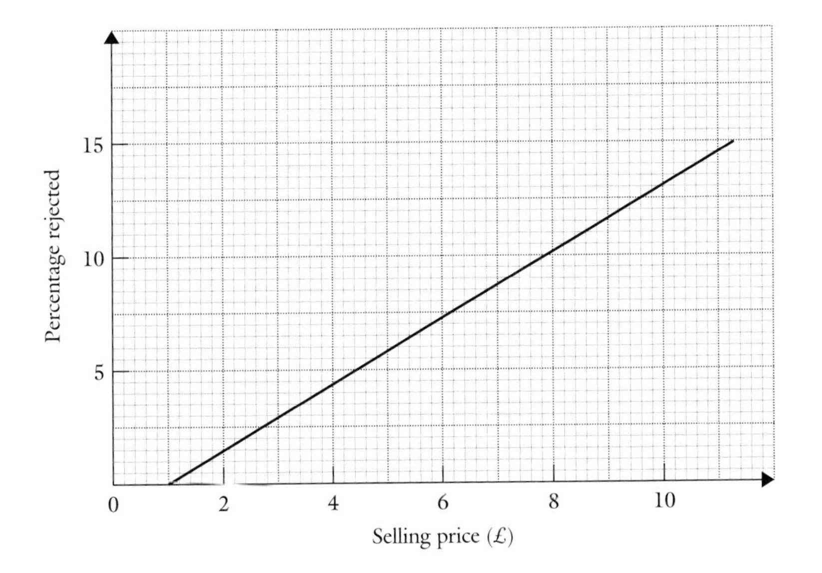


a if the mugs sell at £5 each, what percentage is rejected?

b If 13% of the mugs are rejected, what is the selling price?

2 The graph shows the number of complete patterns in a roll of dress material against the length of material bought.

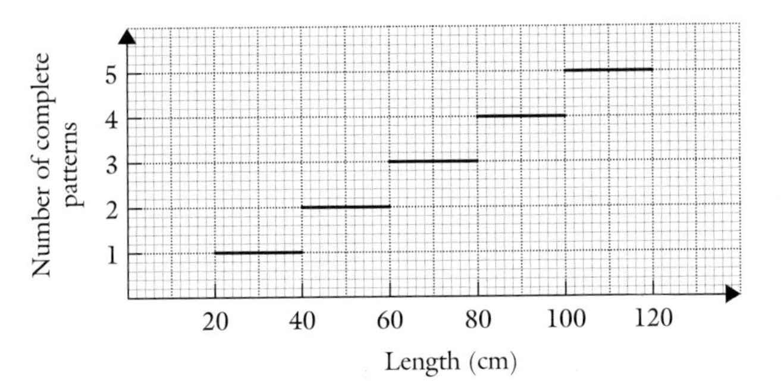


a If 1.2 m of material is bought, how many complete patterns will there be?

b A skirt is to be made with 3 complete patterns plus an allowance of 12 cm for hem and waist band.
What is the shortest length of material which could be bought?

5 The length of time for which a chicken should be cooked depends on the weight of the chicken, as shown on the graph below.

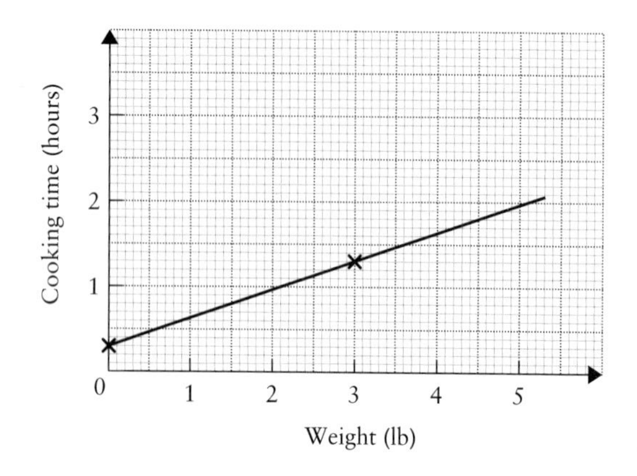


a For how long should a chicken weighing 4 lb be cooked?

b A chicken weighing $2\frac{3}{4}$ lb was cooked for 50 minutes. Was this long enough?

c Why is it important to cook chickens for the correct length of time?

6 A pharmaceutical company delivers medicines to chemists. The distance travelled by the delivery vans against the time taken was recorded and the results plotted on a graph.

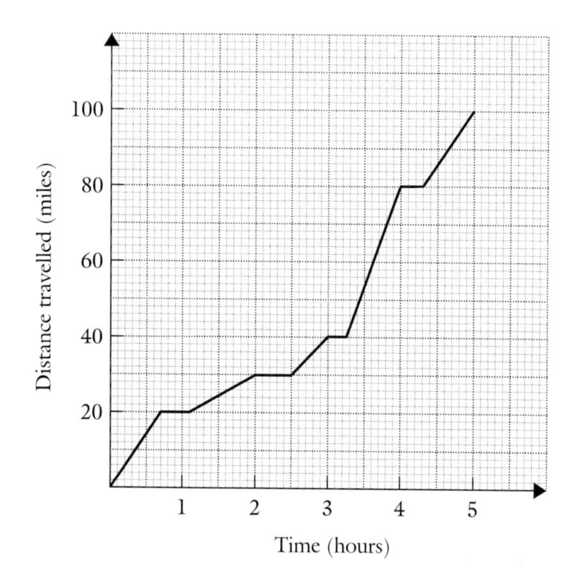


a If the time taken was 2 hours, how far had the van travelled?

b One driver had to travel a distance of 70 miles. How long did the journey take?

## 21.3 Graph plotting

When plotting graphs a certain procedure should be followed.

1 Choose the scales.
2 Draw the axes.
3 Scale the axes.
4 Plot the points.
5 Join the points.

The information from which the graph is to be drawn is usually given in a table.

**EXAMPLE**

£1000 is invested in a savings account for 5 years. The amount in the account at the end of each year is shown on the table:

| Number of years ($x$) | 0 | 1 | 2 | 3 | 4 | 5 |
|---|---|---|---|---|---|---|
| Amount in £ ($y$) | 1000 | 1100 | 1210 | 1331 | 1464 | 1610 |

Plot the graph and find

**a** the amount in the account after $2\frac{1}{2}$ years

**b** the length of time for which the money must be invested to amount to £1500.

**1** Choose the scales:

(i) Find the range of values for each axis.

On the horizontal axis the range is 5.
On the vertical axis the values range from 1000 to 1610, i.e. the range is 610.

(ii) Count the number of large squares in each direction on the graph paper.

The graph paper provided here has 18 large squares in both directions and a grid with five small divisions in each large division.

The aim is to draw as large a graph as possible, but the scale must be easy to read.

Taking the 18 large division width for the horizontal axis, the range of 5 does not divide exactly into 18.

A convenient scale would be 2 large divisions representing 1 year. On the vertical axis, a range of 610 does not divide into 18 large divisions, but 14 cm is divisible by 700. There will be little wastage of space if a scale of 2 large divisions representing £100 is chosen. The axis is scaled from £1000 to £1700.

**2** Draw the axes.

Once the scales have been chosen, there is no problem in placing the axes.

The origin can be placed 2 large divisions from the left and 2 large divisions from the bottom edges of the graph paper.

This allows space for labelling the axes. Label the origin O.

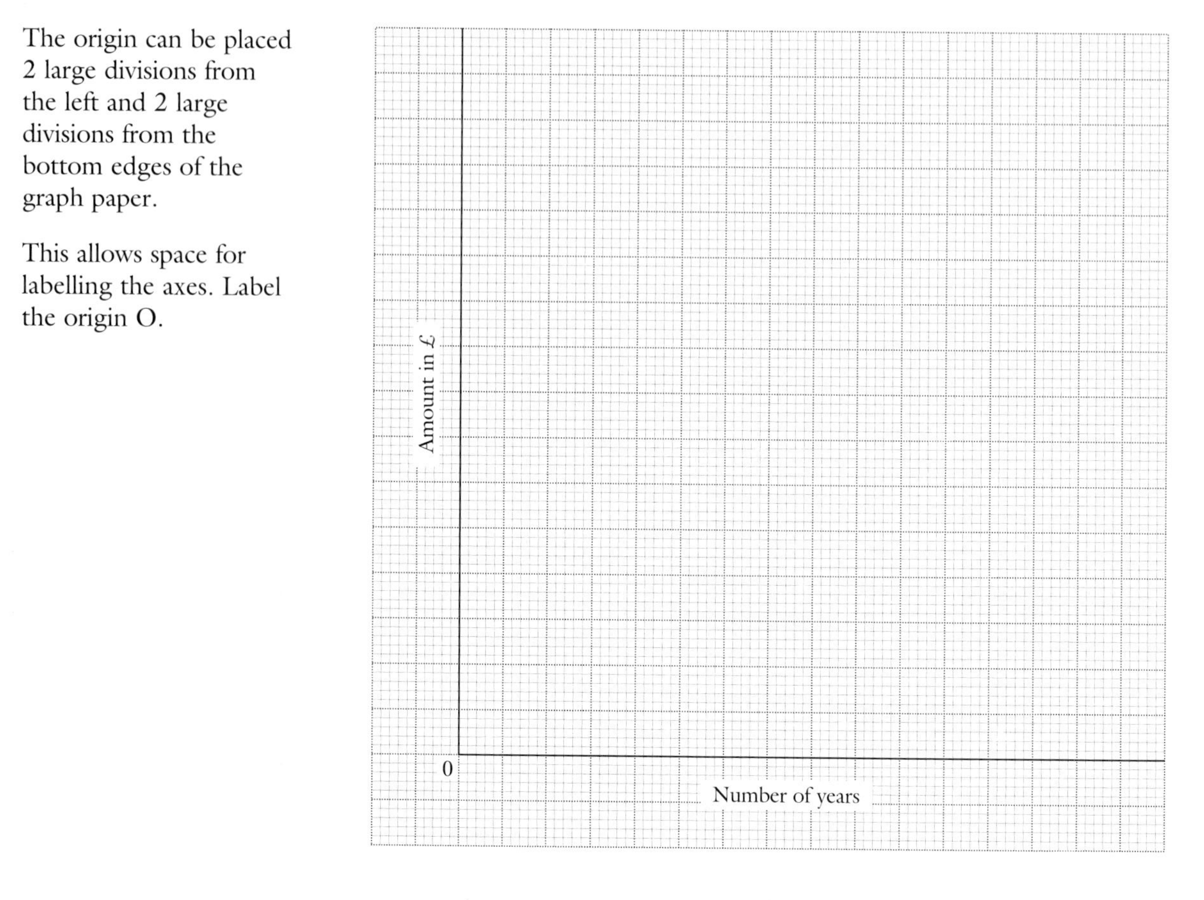


**3** Scale the axes:

Mark the position of each year along the horizontal axis as shown on the following graph.

Mark the position of each £100 along the vertical axis as shown.

**4** Plot the points.

Plot the six points given as explained on page 115.

**5** Join the points.

The points are joined with a **smooth** curve.

Trace through the points with a pencil, but without touching the paper, to find the shape of the curve. When you are satisfied that the path of the curve is smooth, draw the curve through the points in one movement.

If the values of the coordinates have been rounded, the line may not pass exactly through each point.

6 To find the required information draw appropriate lines on the graph (see page 115) and read off the values:

a £1270 b 4.3 years

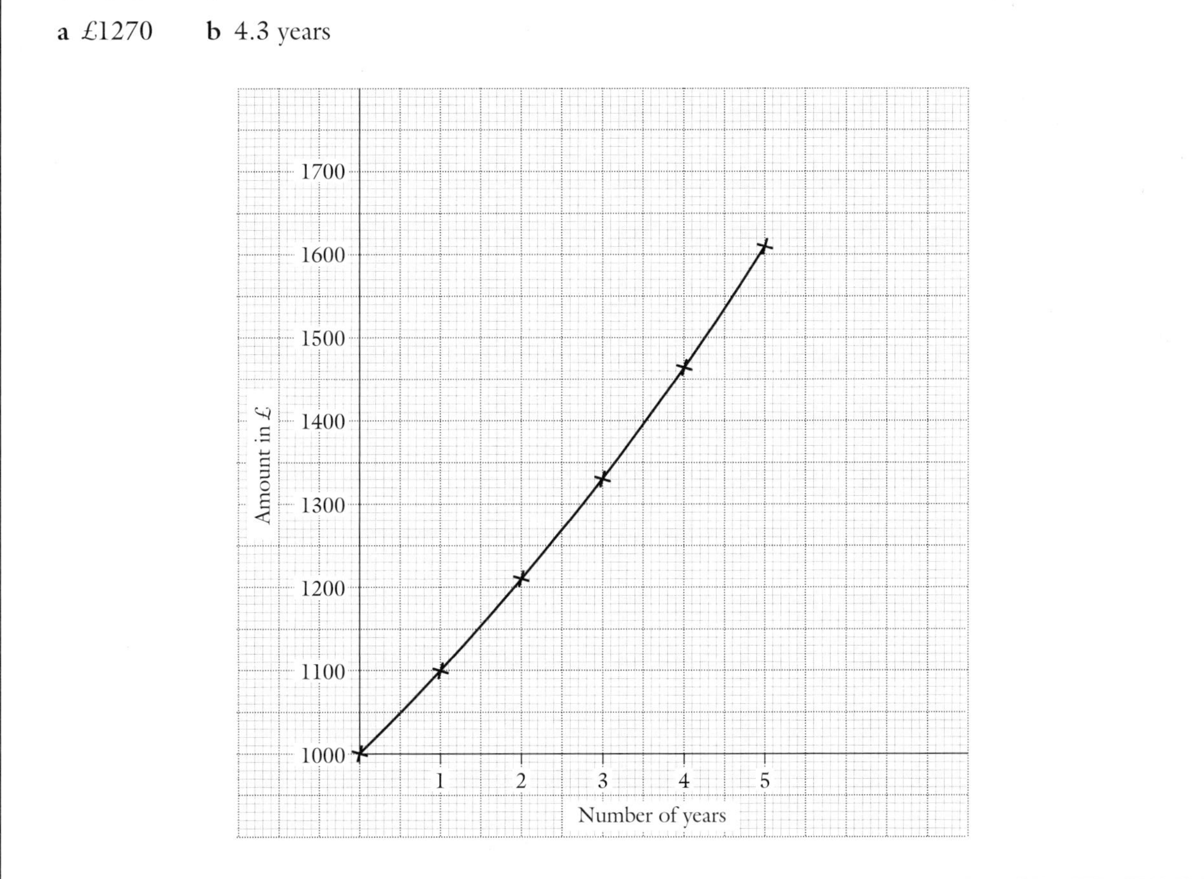


EXERCISE 21.2

In the following questions, you are given the information connecting the two variables in a table. In each case plot the points and join them with a smooth curve.

1 Mrs Lister weighed herself now and again through the year. After one Christmas she found her weight was over $10\frac{1}{2}$ stone.

She joined Weightwatchers and was determined to reach her target weight of $8\frac{1}{2}$ stone before the next Christmas.

The table shows her weight in stones and pounds at the beginning of each given month (1 stone = 14 lb.)

| Month/year | Jan/98 | May/98 | Sep/98 | Jan/99 | May/99 | Sep/99 |
|---|---|---|---|---|---|---|
| Weight in st–lb | 9–2 | 9–12$\frac{1}{2}$ | 10–6 | 10–8 | 10–4 | 9–6 |

a What weight was Mrs Lister in August 1998?

b What was the heaviest weight she reached?

c During which months did she weigh 9 stone 7 lb?

d Do you think she would have reached her target weight before Christmas 1999?

2 While at a party, William spills a glass of red wine over the table cloth.

The stain spreads rapidly in a circular shape and the table shows how the area increases as the diameter of the stain increases.

| Diameter (cm) | 0 | 2 | 4 | 6 | 8 | 10 |
|---|---|---|---|---|---|---|
| Area ($cm^2$) | 0 | 3 | 12 | 27 | 48 | 75 |

a What area does the stain cover when the diameter is 9 cm?

b What is the radius of the stain when its area is 50 sq cm?

c The stain stops spreading when its diameter is 10.9 cm. What area does the stain eventually cover?

3 The table shows the average number of hours the sun shone during a day in the middle of each month, recorded at a Spanish resort.

| | Jun | Jul | Aug | Sep | Oct | Nov | Dec | Jan | Feb | Mar | Apr | May | Jun |
|---|---|---|---|---|---|---|---|---|---|---|---|---|---|
| Hours of sunshine | 16 | 15.5 | 14 | 12 | 10 | 8.5 | 8 | 8.5 | 10 | 12 | 14 | 15.5 | 16 |

a During which months might you expect an average of about $10\frac{1}{2}$ hours of sunshine in a day?

b What were the average hours of sunshine at the beginning and end of May?

c Which monthly periods had the smallest increase in hours of sunshine?

4 One day when Heather is having a bath, she notices that the water level is gradually decreasing and realises that the plug is leaking. When she started her bath the depth of water was 300 mm. Plot the points given in the table below to show the depth of water remaining at a given time.

| Time (in minutes) | 0 | 1 | 2 | 3 | 4 | 5 |
|---|---|---|---|---|---|---|
| Depth of water (in mm) | 300 | 110 | 41 | 15 | 5.5 | 2.0 |

a What was the depth of the water after $2\frac{1}{2}$ min?

b After how long was the bath half empty?

c What was the drop in water level after $1\frac{1}{2}$ min?

# Solutions to Exercises

*Note.* Solutions are given for exercises with a 'closed' numerical answer. They are not provided for 'open-ended' questions and for some questions leading purely to an illustration (e.g. a pie-chart or a graph).
Numerical answers are generally given correct to 3 significant figures.

*Exercise 1.1*

**1** **a** 5 tens **b** 7 hundreds **c** 3 thousands
**d** 4 units **e** 6 tenths **f** 2 hundredths
**2** **a** 10, 17, 21, 45, 54, 86
**b** 14, 23, 32, 104, 203, 230
**c** 6, 60, 61, 600, 601, 610
**d** 11, 99, 101, 110, 999, 1001
**e** 297, 300, 399, 400, 420
**3** 7510 **4** 2369
**5** **a** 430 **d** 132 **g** 516
**b** 16 700 **e** 590 **h** 7030
**c** 20 000 **f** 935.8 **i** 130
**6** **a** 32 **d** 40.3 **g** 0.0027
**b** 3 **e** 0.65 **h** 0.051
**c** 40 **f** 0.065 **i** 0.0002
**7** **a** 100 **b** 400 **c** 4100 **d** 6200

*Exercise 1.2*

**1** 41.9 **2** 180.3 **3** 133.97 **4** 16.7 **5** 16.71
**6** 51.75 **7** 169.4 **8** 702.36 **9** 1770 **10** 7.8
**11** 72.31 **12** 47.86

*Exercise 1.3*

**1** 26 **2** 36 **3** 5 **4** 1 **5** 1 **6** 7
**7** 0 **8** 66 **9** 133 **10** 2 **11** £4.96 **12** 4
**13** £78.01 **14** **a** 988 **b** £4755
**15** **a** 2 h 7 min (127 min) **b** £19.05
**16** 7 packs, £1.35 **17** **a** 47.5 miles **b** £12.92
**18** £1183.80 **19** £160.50
**20** **a** 7.39 kg **b** 135 **21** **a** 9 **b** 14.6 cm

*Exercise 1.4*

**1** 2 **b** −4 **c** −10 **d** 4 **e** 6 **f** 40 **g** 0 **h** −49
**2** **a** −1 **b** 17 **c** −13 **d** 20 **e** −3 **f** 0 **g** 50 **h** 36

*Exercise 1.5*

**1** **a** −18 **b** −3 **c** 28 **d** $1\frac{3}{4}$ **e** −15
**f** −27 **g** 8 **h** 36 **i** −1 **j** −10

*Exercise 1.6*

**1** 1, 4, 9, 16, 25, 36, 49, 64, 81, 100
**2** 1, 8, 27, 64, 125, 216
**3** **a** 5 **b** 11 **c** 25 **d** 15 **e** 14
**4** **a** (i) 3 (ii) −4 (iii) 9
**b** 2 **c** (i) 3 (ii) 5
**5** $1 + 3 + 5 + 7 + 9$
**6** 8 **7** 27 **8** **a** 2 **b** 3 **c** 6 **d** $\frac{1}{2}$ **e** −4
**9** 3 **10** 100

*Exercise 2.1*

**1** 9.74 **2** 0.36 **3** 147.5 **4** 29
**5** 0.53 **6** 4.20 **7** 1250 **8** 0.004
**9** 270 **10** 460.0

*Exercise 2.2*

**1** **a** 3500 **b** 240 **c** 1200 **d** 300 **e** 250
**f** 10 **g** 40 **h** 5
**2** **a** 9 **b** 0.06 **c** 24
**3** Incorrect calculations are **c**, **d**, **e**, **f**, **h**, **i**.

*Exercise 2.3*

**1** 52p **2** 6 tins **3** 5 **4** **a** 27p **b** 3p
**5** 2m **6** 25p **7** £10.91 **8** £13.86
**9** £1458.90 **10** 26

*Exercise 2.4*

**1** 11.8 **2** 73 700 **3** 2.70 **4** 0.173 **5** 88.9
**6** 29.0 **7** 21.4 **8** 26.0 **9** 0.786 **10** 882.65
**11** 0.829 **12** 11 **13** 3.3

*Exercise 3.1*

**1** **a** 8 **b** 8 **c** 18 **d** 9 **e** 7 **f** 9
**2** **a** $\frac{2}{3}$ **b** $\frac{3}{4}$ **c** $\frac{2}{9}$ **d** $\frac{1}{3}$ **e** $\frac{3}{4}$ **f** $\frac{2}{3}$ **g** $\frac{6}{13}$ **h** $\frac{2}{3}$
**3** $\frac{2}{3}$ **4** **a** $\frac{9}{16}$ **b** $\frac{3}{16}$ **5** $\frac{3}{16}$ **6** $\frac{1}{4}$ **7** $\frac{4}{7}$
**8** **a** $\frac{20}{9}$ **b** $\frac{31}{6}$ **c** $\frac{59}{12}$ **d** $\frac{111}{10}$ **e** $\frac{35}{4}$ **f** $\frac{38}{3}$ **g** $\frac{143}{20}$ **h** $\frac{143}{7}$
**9** **a** $1\frac{2}{7}$ **b** $7\frac{1}{5}$ **c** $3\frac{1}{3}$ **d** $7\frac{5}{8}$ **e** $13\frac{10}{11}$ **f** $10\frac{7}{12}$
**g** $8\frac{4}{5}$ **h** $23\frac{1}{2}$

*Exercise 3.2*

**1** $1\frac{3}{20}$ **2** $\frac{1}{24}$ **3** $\frac{2}{3}$ **4** $\frac{3}{4}$ **5** $\frac{3}{20}$
**6** **a** 6 **b** (i) $\frac{1}{4}$in (ii) $\frac{11}{16}$in **7** $\frac{1}{12}$
**8** $\frac{2}{15}$ **9** **a** $\frac{1}{12}$ **b** 3
**10** **a** $\frac{3}{25}$ **b** 400 mg aspirin, 40 mg ascorbic acid, 60 mg caffeine
**11** $3\frac{5}{12}$ **12** 7 in **13** $\frac{3}{16}$ in **14** $1\frac{7}{8}$ in

*Exercise 3.3*

**1** **a** $\frac{8}{12} = \frac{2}{3}$, $0.\dot{6}$, **b** $\frac{1}{4}$, 0.25 **c** $\frac{5}{8}$, 0.625
**2** **a** 0.1 **b** 0.5 **c** 0.75 **d** 1.45
**e** 4.84 **f** $2.8\dot{3}$ **g** $7.\dot{4}$ **h** $3.\dot{1}4285\dot{7}$
**3** **a** $\frac{1}{2}$ **b** $\frac{1}{4}$ **c** $\frac{2}{3}$ **d** $1\frac{1}{3}$ **e** $1\frac{3}{10}$
**f** $2\frac{4}{5}$ **g** $3\frac{3}{5}$ **h** $2\frac{3}{20}$ **i** $\frac{1}{8}$ **j** $\frac{3}{8}$
**5** **a** Insert d.p. into numerator 1 place from right.
**b** Multiply numerator by 2 and change to tenths.
**c** Insert d.p. into numerator 2 places from right.
**d** Change to tenths then divide by 2.
**e** Multiply numerator by 4 and change to hundredths.
**6** 34 **7** **a** (i) $1\frac{2}{3}$ (ii) 1.67 **b** 80p
**8** **a** 20 sheets **b** (i) $20\frac{5}{6}$ (ii) 20.83
**9** (i) $4\frac{2}{3}$ hours (ii) 4.67 hours
**10** (i) $3\frac{1}{5}$ hours (ii) 3.20 hours **11** (i) $1\frac{3}{4}$ (ii) 1.75

*Exercise 4.1*

**1** 5 : 6 **2** 4 : 5 : 10 **3** 20 : 9
**4** **a** 3 : 2 **b** 10 : 1 **c** 12 : 1 **d** 1 : 4 **e** 4 : 3
**f** 3 : 2 **g** 3 : 8 **h** 3 : 2 **i** 2 : 3
**5** **a** 8 **b** 2 **c** 1.2 **d** 0.6 **e** 3.2 **f** 2.6
**6** 410 and 1025 **7** $13\frac{1}{2}$ kg **8** 3 : 20
**9** **a** 125 **b** 200 **10** 3 : 16 **11** 28 fl oz
**12** 1 : 8 **13** **a** 720 **b** 480 **14** 40 : 1 **15** 66

*Exercise 4.2*

**1** £260, £390 **2** £1500, £4500, £6000
**3** £48, £32, £24 **4** £1750, £1250, £1000
**5** 1360 g, 240 g **6** 20
**7** **a** £6318, £4212 **b** £15 760 **c** £9924
**8** £26 154, £34 872, £17 436, £26 154
**9** 27, 45 **10** 40
**11** **a** (i) 5 : 3 (ii) £7.50, £4.50 **b** 7.33, £4.67
**12** £24 **13** 1813 **14** **a** 192 **b** 720

*Exercise 4.3*

**1** £3.18 **2** £2.66 **3** £1.26 **4** £1.84 **5** £4.40
**6** £242 **7** £430.50 **8** 44 **9** 200 **10** 4 hours
**11** £1845 **12** 84 **13** $\frac{11}{16}$ in

*Exercise 5.1*

**1** **a** 20% **b** $12\frac{1}{2}$% **c** 70% **d** 65% **e** $66\frac{2}{3}$% **f** 36%
**g** 175% **h** 250%
**2** **a** $\frac{3}{5}$ **b** $\frac{1}{4}$ **c** $\frac{1}{10}$ **d** $\frac{17}{20}$ **e** $\frac{3}{20}$ **f** $1\frac{3}{20}$ **g** $\frac{3}{8}$ **h** $\frac{1}{3}$

**3**

| | Fraction | Decimal | Percentage |
|---|---|---|---|
| a | | 0.75 | 75% |
| b | $\frac{1}{2}$ | | 50% |
| c | | 0.125 | $12\frac{1}{2}$% |
| d | $\frac{1}{3}$ | $0.\dot{3}$ | |
| e | $\frac{3}{8}$ | | $37\frac{1}{2}$% |
| f | | 0.7 | 70% |
| g | $\frac{7}{20}$ | 0.35 | |
| h | $\frac{2}{3}$ | | $66\frac{2}{3}$% |
| i | | 0.6 | 60% |
| j | $\frac{5}{8}$ | 0.625 | |

*Exercise 5.2*

**1** **a** £1.38 **b** £318.62 **c** £22.12 **d** £13.91 **e** £5.17
**f** £8.73 **g** £97.20 **h** £44.20

*Exercise 5.3*

**1** **a** £10.12 **b** £4.07 **c** £5.50 **d** £10.79 **e** £10.04
**2** **a** £108.18 **b** £1.05 **c** £792.00 **d** £233.75
**e** £156.00 **f** £21.69

*Exercise 5.4*

**1** **a** £24.00 **b** £8.40 **c** £14.08 **d** 36p **e** £10.39
**2** **a** £49.77 **b** 59p **c** £51.20 **d** £23.76 **e** £59.43
**f** £1.09

*Exercise 5.5*

**1** **a** 80.0% **b** 3.45% **c** 175% **d** 50% **e** 54.5% **f** 130%
**2** **a** +23.1% **b** +20.0% **c** +375% **d** −20.0%
**e** −45.5% **f** +464%

*Exercise 5.6*

**1** **a** 18% **2** **a** 260 **b** 78 **3** 12%
**4** **a** 20 000 **b** 7500 **c** 36.2%
**5** **a** 406 **b** 146 **6** £2.64 **7** £3.43
**8** £1335.70 **9** 616 **10** £2240

*Exercise 6.1*

**1** £172.20 **2** £479.67 **3** £6.43 **4** £11582.40
**5** £134.28 **6** £23 700 **7** £7320 **8** £186.82
**9** £5244 **10** £194.36 **11** £5.526 **12** £36.95
**13** **a** £259.70 **b** £155.75 **c** £441 **14** £4.2667

*Exercise 6.2*

**1** £344.10 **2** **a** £4.50 **b** 4 **3** £187.68 **4** £259.89
**5** £8.43 **6** £253.80 **7** £279.90 **8** £218.40
**9** £269.33 **10** £166.50 **11** £266.40
**12** Andrews £204.18
Collins £268.92
Hammond £263.94
Jali £318.72
Longman £308.76
**13** £5.45

*Exercise 6.3*

**1** £1560 **2** £24.72 **3** £1883.20 **4** £493.20
**5** £1138 **6** £1456
**7** **a** Firm B by £500 **b** Firm A by £250
**8** £283.55 **9** £94 **10** £60 **11** £99.40
**12** £304 **13** £380.70

*Exercise 6.4*

**1** £59.04 **2** £56.40 **3** £3.99 **4** £228.70
**5** £182 **6** £362 **7** £243.50 **8** £31 **9** £440
**10** 1570 Fr **11** £62.08 **12** £98.20

*Exercise 6.5*

**1** **a** (i) £4385 (ii) £115 (iii) £11.50
**b** (i) £4385 (ii) £565 (iii) £56.50
**c** (i) £4385 (ii) Nil (iii) Nil
**2** £132 **3** £9.31 **4** **a** £85.80 **b** £106.50
**5** £10.76 **6** £136.50 **7** £9.96
**8** **a** (i) £4385 (ii) £815 (iii) £81.50
**b** (i) £4385 (ii) £1387 (iii) £138.70
**9** £9.70 **10** £9.35 **11** £2.57 **12** £147.50 **13** £146

*Exercise 6.6*

**1** **a** £1503.90 **b** £934.98 **c** £2429.20
**2** **a** £2935.20 **b** £2935.20
**3** **a** £61.45 **b** £35.59 **c** £96.60 **4** £5186.70
**5** £370.08 (or .07) **6** £5593.04 **7** £391.71
**8** £206.91 **9** £331.21 **10** £4759.90
**11** £389.88 (or .87) **12** £239.41 **13** £5704.14

*Exercise 6.7*

**1 a** (i) £7751.60 (ii) £645.97
**b** (i) £8271.60 (ii) £689.30
**c** (i) £7129.60 (ii) £594.13
**2 a** £8151.60 **b** £788.50
**3** £16 543.20 **4 a** £6481 **b** Yes, by £648.60

*Exercise 6.8*

**1 a** £3.67 **b** £25.74 **c** £53.93 **d** £8.37 **e** £4.44
**2 a** £31.20 **b** £4232.80
**3 a** £184 **b** £10.80 **c** £143.95

*Exercise 7.1*

**1 a** 137.76 kunas **b** 128.24 lats **c** $152.47
**2 a** £76.34 **b** £4549.18 **c** £3.82
**3** £8.47 **4 a** £80.31, 40.2
**5 a** 3913.20 **b** 291.28
**6 a** 177, 940 **b** 70 270 **c** £201.33 **d** £1.86
**7 a** 771.50 **b** £93.99 **c** £6.01

*Exercise 7.2*

**1 a** £1 **b** £1 **c** £8.01 **2 a** £3 **b** £6

*Exercise 8.1*

**1** £26.60 **2** £40.95 **3** £1.66 **4** 25p **5** £3.36
**6** £105.16
**7** A, £14.81; B, £14.30; C, £14.97; B offers the best deal
**8** £97.49 **9** £7.53, £95.88
**10 a** 59p, £5.54 **b** £17, £357 **c** £2.60, £31.50
**d** £46.92, £322.92 **e** £1.92, £15.66

*Exercise 8.2*

**1** £1.22 **2** £9114.89 **3** £17.98 **4** £1744.19
**5** £106.55 **6 a** £134.03 **b** £765.87
**7 a** £3.11, £25.88 **b** £1.93, £16.06 **c** £5.89, £49.10
**d** £21.43, £178.56 **e** £35.36, £294.63
**f** £12.32, £102.67 **g** £64.28, £535.71
**h** £1.28, £10.71
**8** $4.73 **9** £176.91 **10** £9.84

*Exercise 9.1*

**1 a** (i) £1 (ii) 25%
**b** (i) £7 (ii) 35%
**c** (i) 9p (ii) 12%
**d** (i) £18 (ii) 20%
**e** (i) £100 (ii) 19%
**f** (i) £3.99 (ii) 11.1%
**2** 33.3% **3 a** £4.32 **b** 12% **4** £9719.80
**5** 20% **6** 22.4% **7 a** 95.1% **b** 1.46%

*Exercise 9.2*

**1 a** £35 **b** 17.9% **2 a** £1174.50 **b** £225 **c** 23.7%
**3 a** £28.20 **b** 24.5% **4 a** £3.25 **b** 18.1%
**5 a** £1200 **b** £1518.30 **c** £219.70 **d** £1738
**e** £538; 44.8%
**6 a** £1080 **b** £6.75 **c** £1404 **d** 71.22%

*Exercise 10.1*

**1 a** £252 **b** £82.50 **c** 2 years **d** 5 years
**2** £2295, £9045 **3** £231.25 **4** $7\frac{1}{2}$%

*Exercise 10.2*

**1** £7986 **2** £2315.25 **3** £7504.67, £984.67
**4** £4655.45; £1551.82 **5 a** £69.16 **b** 6.92%
**6** The second paying 8.75% per annum, ease of access to the account/services available/extras, e.g. cheque book, credit card.
**7** £826.69

*Exercise 10.3*

**1 a** £210.70 **b** 5.2675%
**2 a** £176.52 **b** 5.884%
**3 a** £321.66 **b** 6.4332%
**4 a** £399.26 **b** 4.99075%
**5 a** £158.64 **b** 6.3456%
**6 a** £246.71 **b** 6.16775%

*Exercise 11.1*

**1** (i) Year is incorrect – should be 1999.
(ii) Alteration should be initialled.
(iii) Amount in figures should read £43.
(iv) It is usual to add 'only' to an amount in whole pounds and/or draw a line through the remaining space to prevent additions to the amount.
**2** (i) Insufficient funds in the account to cover the cheque.
(ii) The cheque is unsigned.
(iii) The cheque was completed incorrectly.
(iv) The cheque is out of date.
**3** The cheque has 'bounced' and the payee must go back to the drawer (writer of the cheque) for payment.
**4** The cheque must be paid into the bank account of the payee.
**5** The card guarantees that the bank will honour cheques up to the amount on the card.
**6** In the case of theft, one cannot be used by the thief without the other.

*Exercise 11.2*

**1 a** Southford **b** 0123456 **c** 30/3/89 **d** £444.01
**e** £29.65 **f** £215.00 **g** Direct debit **h** £225
**i** The account is in credit. **j** Cheque numbers

*Exercise 12.1*

**1 a** £43.00 **b** £38.70 **2** £26.25
**3 a** £223.44 **b** £33.45 **c** 8.8% per annum **d** 17.6%
**4 a** £5309.74 **b** 19.7%
**5 a** £1849.75 **b** £5549.25 **c** £8809.75 **d** 25.4%
**e** 25.4%
**6 a** £11 138.70, £12 111.90 **b** 17.6%, 15.6%
**7** A because APR values are 28% and 34.9%.
**8 a** (i) £1680 (ii) £9240 **b** £1890
**c** (i) $33\frac{1}{3}$% (ii) APR $\approx$ 22.2%

*Exercise 12.2*

1 £277.20 2 £207.51 3 £187.01
4 £173.90 5 £438.06 6 £419.44

*Exercise 12.3*

1 a (i) £1.52; £1.34 (ii) 500 g size b 250 g size
2 a (i) 4.49 g (ii) 4.65 g b 600 g jar
c possible wastage with larger sizes/sell by date/taste
3 a E15: £5.49; E10: £5.68$\frac{1}{2}$; E3: £6.10
b (i) E15: more economical for a family
(ii) E3: less to pay out of pension, not heavy to carry
(iii) E10: good compromise, but an argument could be made for any size
(iv) E3: small size for limited storage space.

*Exercise 13.1*

1 a 66 865 b 42 735 c 68 138 d 94 238
2 a £94.65 b £136.24 c £99.27 d £104.82
e £33.34
3 a £21.38 4 a 902 b 65 210

*Exercise 13.2*

1 a £168.30 b £118.15 c £164.72 d £190.19
2 388.5 3 £41.49
4 a 148 b 155.4 c £66.82 d £76.22
5 a (i) 5839 (ii) 211
b (i) 221.55 (ii) £88.18 (iii) £96.88

*Exercise 13.3*

1 £39.29 2 £37.19 3 a £31.77 b £37.33
4 a £24.96 b 169 c £7.10 d £32.06 e £5.61
f £37.67

*Exercise 14.1*

1 a £25 b £54 c £1477.33 d £222 e £45
2 a £60 b £34 c £1.01 d £1200 e £85
3 £180 4 £215.20 5 £7180 6 £2.66
7 £9500 8 £49.93

*Exercise 14.2*

1 50% 2 21% 3 10$\frac{1}{2}$% 4 9.6% 5 6.5% 6 4.4%

*Exercise 14.3*

1 £2304 2 £630 3 £328.05 4 £2414.45

*Exercise 14.4*

1 40% 2 62.5% 3 26% 4 27.5% 5 128% 6 22.5%

*Exercise 16.1*

(Frequencies only are given)
1 8, 7, 5, 5, 4, 3, 1, 2, 1, 1, 1, 0, 1, 0, 1
2 1, 3, 5, 10, 19, 6, 6
3 18, 17, 13, 7, 4, 2, 1, 0, 1
4 4, 8, 9, 8, 9, 6, 2, 2, 2
5 a 1, 4, 4, 4, 8, 5, 4, 3, 5, 1, 1
b 1, 8, 15, 8, 6, 2
6 1, 14, 23, 19, 6

*Exercise 17.1*

1 a East Lynne b 50 c 410
4 a 175 b 8.3% c 4 : 1
7 a Soft white baps b 6 c 61.9(62)%

*Exercise 17.2*

1 a Oak
b Oak 80; Elm 10; Chestnut 20; Beech 40; Conifer 70; Cedar 25
c 245
6 a A; Adverts on hoarding B are aimed at more prosperous suburban inhabitants.
b 2
c Football club and health education

*Exercise 17.3*

1 a 160 b 120 c 220 d 140
6 a 40° b 25 c 50 d $\frac{5}{18}$
9 a 325 b 290 c 210, 87.8%
11 a 72.2% b 458 300 acres

*Exercise 17.4*

5 a 1 040 000 b 1927, 1951, 1974 c 1940
6 a Tiredness as the week goes by b 226
7 a £180
b £209
c There are more points plotted around 1993 which fit the straight line.
d 1990

*Exercise 17.5*

2 b A low absence rate is more likely on Mondays and a high one on Fridays.
3 b Book 1: Science fiction.
Book 2: Child's story.
Book 1 has a greater number of long words and Book 2 has a greater number of short words.
4 There has been a move to 'Traditional Style' from 'Modern Design'.
6 There are more women than men over the age of 45.

*Exercise 18.2*

1 a Weak positive
b High negative
c None
2 Weak positive
3 Quite high negative
4 Quite high positive
5 a Positive b None c Negative d Positive
6 b No
c Over a period of time both variables have changed, but this is a coincidence. The smaller cod catches are probably due to depleted fish stocks. The percentage vaccinated refers to an independent event.

*Exercise 18.3*

The equations of the lines are given to enable students to check the position of their lines. Answers given are by calculation, answers from graphs should be correct to two s.f.

**1 a** $\bar{x} = 25, \bar{y} = 2.09$
**c** $y = -0.113x + 4.91$
**d** (i) 25.8 min (ii) 16.9 min
**2 a** (i) 64 kg (ii) 73.8 kg
**b** (ii) $y = -48.8 + 0.662x$
(iii) 60.4 kg
**3 a** $y = 13.1 + 2.28x$
**b** 79 marks
**4 a** $y = 52.9 - 0.781x$
**b** 39.6 kcal/hour/$m^2$
**c** 33.3 kcal/hour/$m^2$
**d** The relationship is not linear outside the given range.
**5 b** 14 500
**6 b** (i) About 5.4 years (ii) About 31 days
**7 b** The greater the number of storks, the greater the number of babies born (or vice versa).
**8 a** Reported cases of illness increase as the population increases (i.e. a positive correlation).
**c** (i)300 (ii) 350
**d** actual figures are considerably lower than the predictions.

*Exercise 19.1*

**1** $\frac{5}{8}$ **2** 0 **3** $\frac{1}{3}$ **4** 0
**5 a** Statistical evidence
**b** Subjective estimate
**c** Statistical evidence
**d** Statistical evidence
**e** Subjective estimate
**6 a** $\frac{3}{5}$ or 0.6 **b** $\frac{57}{100}$ or 0.57
**c** greater amount of data used.

*Exercise 20.1*

**1** $7x + 5y$ **2** $2p - q$ or $q - 2p$ **3** $\dfrac{2a}{b}$
**4** $5x - 8$ **5** $\dfrac{y}{4} + 6$ **6** $4a + 3c$
**7** $4b + 6r + 2w$ **8** $4r + 3p + 2t$
**9** $6l + 8p + 2i$ **10** $22n + 3p$
**11** $4t + 5r + 3s$ **12** $3a + 5c$
**13** $7f + 5s + 3e$ **14** $2h + 3i + 22s$
**15** $28f + 12m + 15w$

*Exercise 20.2*

**1** $5x$ **2** $7n$ **3** $8a$ **4** $2y$
**5** $7a + 6b + 4c$ **6** $x + 5y + 9z$
**7** $9p + 2q + r$ **8** $5x - 3y + z$
**9** $-a - b + 4c$ **10** $3x + 5y$

*Exercise 20.3*

**1 a** 11 **b** −1 **c** 13 **d** 6 **e** $\frac{1}{2}$ **f** 2 **g** −30 **h** −4
**i** 0 **j** 17 **k** −1 **l** −2
**2 a** 1 **b** −9 **c** −5 **d** −6 **e** 0.3 **f** −14 **g** 6
**h** 8 **i** 10 **j** −5 **k** 1 **l** −2
**3 a** 9 **b** −24 **c** 8 **d** $\frac{4}{3}$ **e** −5 **f** $-\frac{3}{4}$ **g** 38
**h** 7 **i** 28 **j** 1 **k** 0 **l** 14
**4 a** $\frac{1}{2}$ **b** 1 **c** 2 **d** −4 **e** $2\frac{1}{2}$ **f** $-\frac{1}{4}$ **g** $8\frac{1}{2}$
**h** $\frac{3}{5}$ **i** 3 **j** −1 **k** −8 **l** $4\frac{1}{2}$

*Exercise 20.4*

**1** $5x$ **2** $3mn$ **3** $\dfrac{2y}{z}$ **4** $abc$ **5** $6pqr$ **6** $\dfrac{4bc}{d}$

**7** $6xy$ **9** $\dfrac{-2p}{q}$ **9** $\dfrac{9xy}{z}$ **10** $a^3$ **11** $a^2b^2$ **12** $3a^3$

**13** $3a^3b^2$ **14** $-6a^2b^2c$ **15** $6a^2b^4c^2$ **16** $9a^2b$

**17** $-18ab^3$ **18** $-2a$ **19** $x^3y^3$ **20** $\dfrac{-9a^2b}{c}$

*Exercise 20.5*

**1** $x^8$ **2** $x^6$ **3** $x^3$ **4** $x^{-1}$ **5** $x^0 = 1$

**6** $\dfrac{1}{x^2}$ **7** 1 **8** $\dfrac{3x^2}{2^2}$

*Exercise 20.6*

**1 a** £972.41 **b** £7895.59 **c** £2945.36
**2 a** 5.155% **b** 2.305% **c** 20.51% **d** 14.84%
**3 a** £101.28 **b** 47
**4** £142
**5 a** (i) £16.20 (ii) £115.20 **b** 12.3p
**6 a** £137.50 **b** £82.80

*Exercise 21.1*

**1** Rough estimate **a** $5\frac{1}{2}$% **b** £10
**2** 'Exact' **a** 5 **b** 72 cm
**3** 'Exact' **a** $1\frac{1}{2}$ hours **b** £2500
**4** 'Exact' **a** £2.19 (£2.20) **b** $1\frac{1}{4}$ min
**5** 'Exact' **a** 1 h 38 min (± 2 min)
**b** No, should be cooked for at least 1 h 12 min
**c** To ensure that all bacteria are destroyed and prevent food poisoning.
**6** Rough estimate **a** 30 miles
**b** 3.8 hours (= 3 h 48 min)

*Exercise 21.2*

**1 a** 10 st 4 lb (±1 lb) **b** 10 st 8 lb
**c** February 1990 and August 1991 **d** Yes
**2 a** 60 $cm^2$ **b** 4.1 $cm^2$ **c** 90 $cm^2$
**3 a** October and February **b** 14.8 and 15.8 hr
**c** December to January and May to June
**4 a** 25 mm **b** 0.7 minutes **c** 235 mm

# Index